Sewage Treatment
in
Hot Climates

A haul of carp from a sewage-fed pond in India [Courtesy of the Director, Central Inland Fisheries Research Institute, Barrackpore]

Sewage Treatment
in
Hot Climates

Duncan Mara
Department of Civil Engineering
University of Dundee
Scotland

A Wiley–Interscience Publication

JOHN WILEY & SONS
London · New York · Sydney · Toronto

Library of Congress Cataloging in Publication Data:

Mara, David Duncan.
  Sewage treatment in hot climates.

  'A Wiley–Interscience publication.'
Includes bibliographies and index.
  1. Sewage—Purification—Tropical condition. I. Title.
[DNLM: 1. Sewage. 2. Sanitary engineering.
3. Tropical climate. WA785 M298s]
TD745.M35   628'.3'0913   75–23421

ISBN 0 471 56784 1

Photosetting by Thomson Press (India) Limited, New Delhi,
printed by photolithography and bound in Great Britain at
The Pitman Press, Bath

To
Neil and Jacqueline

μέγα βι βλίον μέγα κακόν

Preface

There are many excellent texts on sewage treatment but none written specifically for hot climates. Thus waste stabilization ponds, which are the most important method of sewage treatment in hot climates (or at least potentially so) have never been treated in the detail required in an undergraduate textbook. A further example is waste re-use, now receiving considerable attention in industrialized countries. Yet it has far greater implications for social advancement in tropical developing countries. Engineers who are studying or practising in hot climates need to be aware of the potential benefits of, for example, aquaculture so that they can work with planners and agriculturalists to prepare integrated plans for waste treatment and waste re-use. In contrast they need to know less about 'conventional' treatment.

This book has been written primarily for undergraduate students of civil engineering as it assumes no previous knowledge of sewage treatment (although fundamental concepts of fluid mechanics are taken as known). I hope that practising engineers will find the book of some use, especially those chapters on ponds, re-use and nightsoil collection; the remaining chapters may serve them as a useful revision of what should be familiar material.

I have had many useful discussions with many engineers. At the risk of offending some, I wish to express my indebtedness to three in particular: Richard Truran, of Sir Alexander Gibb and Partners (Africa), Nairobi; Dr Richard Feacham, of the University of Birmingham, England; and Dr Michael McGarry, of the International Development Research Centre, Ottawa.

Duncan Mara
Dundee
September 1975

Acknowledgements

Permission to use copyright material was kindly granted by the following:
American Society of Civil Engineers (Figures 12.1, 12.2 and 13.5, and Table 12.1).
Cambridge University Press (Figure 2.2).
Churchill-Livingstone (Figures 2.1, 2.3, 2.4, 2.7, 3.3, 4.1, 7.2 and A1.2).
Institute of Water Pollution Control (quotation on p. 134–5).
Institution of Civil Engineers (Table 13.2 and quotation on p. 129).
The Controller, Her Majesty's Stationery Office (Figures 2.9, 10.3 and 10.4,
    Tables 2.2 and 8.1, and quotations on pp. 111–2 and 117).
Dr T. H. Y. Tebutt (Figure 1.1).
Society for Applied Bacteriology (Table 2.1).
Ward, Aschcroft and Parkman (Figure 14.3).
World Health Organization (Figure 2.5, Table 1.1 and quotations on pp. 72,
    73 and 129).

Principal Notation

A	area (for ponds, mid-depth area), m^2
a	constant
b	autolysis rate, d^{-1}; bar thickness, mm
c	concentration (of DO, solids), mg/l
c_s	DO saturation concentration, mg/l
D	depth, m
d	diameter, m
F	soluble BOD_5, mg/l
h	height, m
K	biofilter rate constant, m/d (equation 10.2)
K_b	rate constant for first order removal of FC, d^{-1}
K_L	saturation concentration, mg/l (equation 2.4)
k_1	rate constant for first order removal of BOD, d^{-1}
k_2	rate constant for second order removal of BOD, $(mg/l)^{-1}\, d^{-1}$
L	BOD (usually BOD_5), mg/l
M	flow peaking factor
N	numbers (of cells, FC)
P	population
Q, q	flow, m^3/d or m^3/s
R_e	Reynolds number
S	specific surface, m^2/m^3; solids concentration in oxidation ditch, mg/l
s	bar spacing, mm
T	temperature, °C
$\mathbf{T}$	mean generation time, d
t	time, d
t^*	retention time, d
U_s	settling velocity, m/s
V	volume, m^3
W, x	width, m
X	cell concentration, mg/l
Y	yield coefficient (equation 8.3)
y	O_2 uptake, mg/l
α	coefficient of retardation, d^{-1}; ratio of O_2 transfer in tap water and waste
β	ratio of c_s in distilled water and waste

γ	sludge loading factor, d^{-1}
δ	reactor dispersion number
θ	Arrhenius constant
κ	rate constant for first order removal of soluble BOD, d^{-1}
λ_s	BOD_5 surface loading, kg/ha d
λ_r	BOD_5 removal, kg/ha d
λ_v	BOD_5 volumetric loading, g/m^3 d
μ	specific growth rate, d^{-1}; viscosity, N s/m^2
$\hat{\mu}$	maximum specific growth rate
ρ	density, kg/m^3
ϕ	$\sum t^*$, d

Subscripts

e	effluent
i	influent
T	temperature
t	time

Abbreviations

BOD_5	5 d, 20 °C biochemical oxygen demand
BOD_u	ultimate biochemical oxygen demand
COD	chemical oxygen demand
DO	dissolved oxygen
FC	faecal coliforms
SS	suspended solids

Units

Generally SI units have been used. The only unit likely to be misunderstood is that used for conductivity, mS/m (millisiemens per metre): mS/m × 10 = μS/cm. This latter unit, although more familiar, is (strictly) not permissible as centimetres are to be avoided in SI. Reference: R. J. Wells, *Journal of the Institution of Water Engineers and Scientists*, **29**, 46 (1975).

Contents

1

What is Sewage?

1.1 COMPOSITION OF SEWAGE

Sewage is the wastewater of a community. It may be purely domestic in origin or it may contain some industrial or agricultural wastewater as well. Initially we will consider only domestic sewage. This is composed of *human body wastes* (faeces and urine) and *sullage* which is the wastewater resulting from personal washing, laundry, food preparation and the cleaning of kitchen utensils.

Fresh sewage is a grey turbid liquid which has an earthy but inoffensive odour. It contains large floating or suspended solids (such as faeces, rags, plastic containers, maize cobs), smaller suspended solids (such as partially disintegrated faeces, paper, vegetable peel) and very small solids in colloidal (i.e. non-settleable) suspension, as well as pollutants in true solution. It is objectionable in appearance and extremely hazardous in content, mainly because of the number of disease-causing ('pathogenic') organisms it contains. In hot climates sewage can soon lose its content of dissolved oxygen and so become 'stale' or 'septic'. Septic sewage has a most offensive odour, usually of hydrogen sulphide.

Approximate analyses of human faeces and urine are given in Table 1.1

Table 1.1 Composition of human faeces and urine*

	Faeces	Urine
Quantity (wet) per person per day	135–270 g	1·0–1·31 g
Quantity (dry solids) per person per day	35–70 g	50–70 g
Approximate composition (%)		
Moisture	66–80	93–96
Organic matter	88–97	65–85
Nitrogen	5·0–7·0	15–19
Phosphorus (as P_2O_5)	3·0–5·4	2·5–5·0
Potassium (as K_2O)	1·0–2·5	3·0–4·5
Carbon	44–55	11–17
Calcium (as CaO)	4·5	4·5–6·0

*From H. B. Gotaas, *Composting: Sanitary Disposal and Reclamation of Organic Wastes*, World Health Organization, 1956.

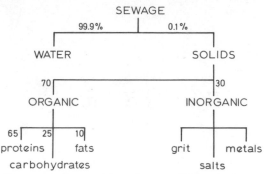

Figure 1.1 Composition of sewage [From T. H. Y. Tebbutt, *Principles of Water Quality Control*, Pergamon, Oxford, 1970]

and, in simpler form, in Figure 1.1. The organic fraction of both is composed principally of proteins, carbohydrates and fats. These compounds, particularly the first two, form an excellent diet for bacteria, the microscopic organisms whose voracious appetite for food is exploited by sanitary engineers in the biological treatment of sewage. In addition to these chemical compounds, faeces and, to a lesser extent, urine contain many millions of intestinal bacteria and small numbers of other organisms. The majority of these are harmless—indeed some are beneficial—but an important minority is able to cause human disease (see Table 2.1).

Sullage contributes a wide variety of chemicals—detergents, soaps, fats and greases of various kinds, pesticides, anything in fact that goes down the kitchen sink and this includes such diverse items as sour milk, vegetable peelings, tea leaves, soil (arising from the preparation of vegetables) and sand (used to clean cooking utensils). The number of different chemicals which are found in sewage is so vast that, even if it were possible, it would be meaningless to list them all. For this reason sanitary engineers use special parameters to characterize wastewaters.

1.2 CHARACTERIZATION OF SEWAGE

As will be explained more fully in Chapter 3, wastes are usually treated by supplying them with oxygen so that bacteria can utilize the waste as food. The general equation is:

$$\text{Wastes} + \text{Oxygen} \xrightarrow{\text{bacteria}} \text{Treated waste} + \text{New bacteria}$$

The complex nature of domestic sewage precludes its complete analysis. Since it is comparatively easy to measure the amount of oxygen used by the bacteria as they oxidize the waste, the concentration of organic matter in the waste is expressed in terms of the amount of oxygen required for its oxidation. Thus if, for example, half a gram of oxygen is consumed in the oxidation of each litre of a particular waste, then we say that this waste has an *oxygen demand* of 500 mg/l, by which we mean that the concentration of organic matter in a litre of the waste is such that its oxidation requires 500 mg.

There are basically three ways of expressing the oxygen demand of a waste:
(1) *Theoretical Oxygen Demand.* This is the theoretical amount of oxygen required to oxidize the organic fraction of the waste completely to carbon dioxide and water. Thus from the equation for the total oxidation of, say, glucose:

$$\underset{180}{C_6H_{12}O_6} + \underset{192}{6O_2} \longrightarrow 6CO_2 + 6H_2O$$

we can calculate that the ThOD of a 300 mg/l solution of glucose is $(192/180) \times 300 = 321$ mg/l.

Because sewage is so complex in nature the ThOD cannot be calculated, but in practice it is approximated by the chemical oxygen demand.
(2) *Chemical Oxygen Demand.* This is obtained by oxidizing the waste with a boiling acid dichromate solution. This process oxidizes almost all organic compounds to carbon dioxide and water, the reaction usually proceeding to more than 95 per cent completion. The advantage of COD measurements is that they are obtained very quickly (within 3 h), but they have the disadvantages that they do not give any information on the proportion of the waste that can be oxidized by bacteria nor on the rate at which bio-oxidation may occur.
(3) *Biochemical Oxygen Demand.* This is the amount of oxygen required for the oxidation of a waste by bacteria. It is therefore a measure of the concentration of organic matter in a waste that can be oxidized by bacteria ('biodegraded'). BOD is usually expressed on a 5 d, 20 °C basis, i.e. as the oxygen consumed during oxidation of the waste for 5 d at 20 °C. This is because the 5 d BOD (usually written 'BOD_5') is more easily measured than is the ultimate BOD (BOD_u) which is the oxygen required for the complete bio-oxidation of the waste. (The reason for the seemingly arbitrary choice of 20 °C and 5 d for the measurement of BOD is given in Section 3.4.) The correct concept of BOD is fundamental to sanitary engineering and a more rigorous treatment of BOD and its removal kinetics is given in Chapter 4.

From the foregoing it is apparent that:

$$ThOD > COD > BOD_u > BOD_5$$

There is no general relationship between these various oxygen demands. However for untreated domestic sewage a large number of measurements have indicated the following *approximate* ratios:

$$BOD_5/COD = 0 \cdot 5$$
$$BOD_u/BOD_5 = 1 \cdot 5$$

The presence of industrial or agricultural wastes may alter these ratios considerably.

The BOD and COD tests are described in Appendix 1.

1.3 STRENGTH OF SEWAGE

The higher the concentration of waste matter in a sewage, the 'stronger' it is said to be: sewage strength is most often judged by its BOD_5 or COD (Table 1.2). The strength of the sewage from a community is governed to a very large degree by its water consumption. Thus in USA where the water consumption is high (350–400 l/hd d) the sewage is weak (BOD_5 = 200–250 mg/l), whereas in tropical countries the sewage is strong (BOD_5 = 400–700 mg/l) as the water consumption is much lower (40–100 l/hd d). Typical analyses of several temperate and tropical sewages are given for comparison in Table 1.3.

Table 1.2 Sewage strength in terms of BOD_5 and COD

Strength	BOD_5 (mg/l)	COD (mg/l)
Weak	< 200	< 400
Medium	350	700
Strong	500	1000
Very strong	> 750	> 1500

The other factor determining the strength of domestic sewage is the BOD (= amount of organic waste) produced per person per day. This varies from country to country and the differences are largely due to differences in the quantity and quality of sullage rather than of body wastes, although varia-

Table 1.3 Analyses of tropical and temperate sewages

Component	Concentration, mg/l						
	Kenya (Nairobi)*	Kenya (Nakuru)+	India (Kodun-gaiyur) ‡	Peru (Lima)²	Israel (Herzliya)‖	USA (Allentown)¶	UK (Yeovil)**
BOD_5	448	940	282	175	285	213	324
SS	550	662	402	196	427	186	321
Total dis-solved solids	503	611	1060	1187	1094	502	—
Chloride	50	62	205	—	163	96	315
Ammonia-cal—N	67	72	30	—	76	12	29

*Average values, 1972.
+24-hour composite sample, 6/7 June 1972.
‡From A. Raman et al., in Low Cost Waste Treatment, CPHERI, Nagpur, 1972.
²From F. Valdez-Zamudio, Science of the Total Environment 2, 406 (1974).
‖From A. Meron et al., Journal of the Water Pollution Control Federation 37, 1657 (1965).
¶From E. W. Steele, Water Supply and Sewage Disposal, McGraw-Hill, New York, 1960.
**Average values, 1969–1972, from M. J. Tarbox, Water Pollution Control 73, 155 (1974).

tions in diet are important. Some figures which have been obtained for the daily per capita BOD_5 contributions are:

Zambia	36 g
Kenya	23 g
S. E. Asia	43 g
India	30–45 g
Rural France	24–34 g
UK	50–59 g
USA	45–78 g

A suitable design value for tropical developing countries is probably about 40 g/hd d (Table 1.4).

Table 1.4 Average breakdown figures for daily per capita BOD_5 contributions (g/hd d)

	USA*	Tropics[+]
Personal washing	9	5
Dishwashing	6	8[§]
Garbage disposal[‡]	31	
Laundry	9	5
Toilet—faeces	11	11
Urine	10	10
paper	2	1[‖]
Total (average adult contribution)	78	40

*From K. Ligman et al., *Journal of the Environmental Engineering Division ASCE*, **100**, 201 (1974).
[+]Conservative estimates.
[‡]Sink-installed garbage grinder.
[§]Includes allowance for food scraps.
[‖]Cleansing material may not be paper —water and leaves are common alternatives.

1.4 WHY TREAT SEWAGE?

Sewage should be treated before its ultimate disposal in a receiving watercourse in order to:

(a) reduce the spread of communicable diseases caused by the pathogenic organisms in the sewage; and
(b) prevent the pollution of surface and ground waters.

These two reasons are interdependent to the extent that a polluted body of water is a potential—and frequently an actual—source of infection, particularly in hot climates. However, there is now an increasing awareness that pollution and contamination of the environment is most undesirable *in itself* and that therefore measures to abate pollution should be judged from an ecological

6

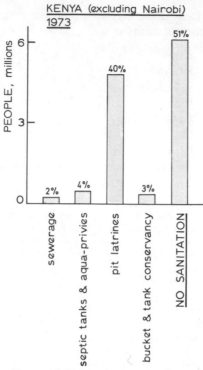

Figure 1.2 Domestic waste disposal facilities in Kenya, 1973 [From information provided by the Water Department, Ministry of Agriculture, Nairobi]

standpoint rather than merely by the improvement they may make to the human condition. Nevertheless, in most tropical developing countries the relative scarcity of funds and the desperate need for sanitary facilities will ensure that, for many years to come, money will be spent primarily on measures directly designed to improve the wellbeing of the people (water supply, sewage disposal, mosquito control etc.) rather than on improving the environment for its own sake. An idea of the extent of the future requirement for sewage disposal facilities can be obtained from Figure 1.2 which shows that half the population of Kenya is without sanitation. Delay in providing an adequate system of sewage collection and disposal, particularly in urban centres, may be excusable—and, unfortunately, inevitable—because of lack of funds, but nevertheless it will have a serious impact on community health, impede social and economic development and thus frequently prevent the best use of scarce resources.

Effluent re-use

One of the best ways to prevent the pollution of surface waters is to use the treated sewage to produce some tangible and beneficial end-product. The principal role for the economic re-use of sewage effluent is in protein production, either as irrigation water or by stimulating the growth of algae and fish (Chapter 13). Domestic sewage is usually considered an actual or potential pollutant, but in many societies it is in fact a scarce resource, a valuable raw material. If sanitary engineers can learn to look at sewage in this light, the pollution problem will, to a very large extent, disappear.

1.5 COLLECTION OF SEWAGE

Sewage is conveyed in pipes, known as sewers, from its place of production to its place of treatment and disposal. Except when sewage is treated in septic tanks which are situated close to the house or houses from where the sewage emanates, the pipework which comprises the reticulation and trunk sewers is usually considerable both in quantity and cost. Costs are usually minimized in hot climates by having a *separate* or dual sewerage system—foul or sanitary sewers which carry only sewage and surface-water drainage channels to carry away stormwater. Large towns and cities may have a *partially separate* sewerage system under which the high-value commercial buildings in the centre of the town are protected from floods by allowing the stormwater falling on this area to enter the sewers. The hydraulic design of sanitary sewers is discussed in Appendix 3.

Non-waterborne sewage

A waterborne sewerage system is undoubtedly best. But it is highly capital intensive and many communities in need of sewerage cannot afford to install a full network of reticulation and trunk sewers. In these cases a nightsoil collection system is necessary; and with proper control it need not be unhygienic. Some modern methods for the collection and treatment of nightsoil are described in Chapter 14.

1.6 FURTHER READING

Metcalf & Eddy, Inc., *Wastewater Engineering: Collection, Treatment and Disposal*, McGraw-Hill, New York, 1972.

2

Essential Microbiology

2.1 Importance of Microbes

The micro-organisms upon whose activities the operation of biological waste treatment processes depends are the *bacteria*, *algae* and *protozoa*. These microscopical organisms belong to the class of organisms known as 'protists' which many authorities consider to form a separate kingdom ancestral to both the plant and animal kingdoms (Figure 2.1).

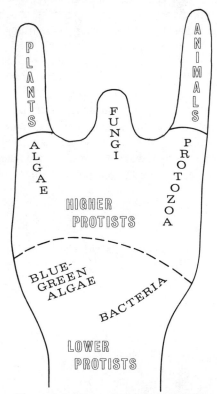

Figure 2.1 The probable relationship between Protists, Plants and Animals

Some bacteria and a few protozoa are human pathogens, as are many *viruses*. Control of their occurence in the environment is therefore of the utmost important and efficient sewage treatment can achieve much towards this end. Sewage treatment is also extremely effective in reducing the incidence of diseases due to worms and other intestinal parasites. A sanitary engineer must have a reasonably good knowledge of faecal bacteria so that he can design sewage treatment works in which they are removed to as high a degree as is practicable. Only in this way is he able to minimize the risks to public health which result from the uncontrolled disposal of untreated or insufficiently treated sewage.

Microbes, however, have a more positive role to play in sewage treatment. Bacteria are the primary degraders of organic wastes: biological waste treatment plants should therefore be designed to enable them to grow—and therefore the waste to be oxidized—at their maximum rate. A knowledge of the rate at which bacteria can oxidize wastes is thus of direct practical importance in the design of sewage treatment works (Chapter 4).

2.2 BACTERIA

Most bacteria are non-photosynthetic single-celled organisms which multiply by splitting into two daughter cells (Figure 2.2). Bacteria come in various shapes and sizes (Figure 2.3) although the rod-shaped ones ('bacilli') are the most common in sewage. Most obtain their energy for growth by the oxidation of organic compounds, but a few are able to use inorganic compounds for this purpose. Organic compounds also serve for most bacteria as sources of carbon which is used to synthesize new cells during growth, although again a few—fortunately of less importance in sewage treatment—use inorganic carbon (i.e. carbon dioxide) for this purpose.

Air—or more correctly, the oxygen in it—is an important environmental factor for bacteria. Air is essential to the 'obligately aerobic' bacteria but toxic to the 'obligately anaerobic' bacteria. However most bacteria are indifferent to its presence or absence, although growth is better in its presence; these are called the 'facultative' bacteria which therefore grow aerobically in the presence of air and anaerobically in its absence.

Temperature is an important factor controlling the rate at which growth occurs. For any one kind of bacterium there is a range of temperature in which growth occurs and, within this range, there is an optimum temperature at which maximum growth occurs (Figure 2.4). Bacteria are usually classified in one of the following three groups according to the range in which their optimum temperature occurs:

Psychrophils $< 20\,°C$
Mesophils $20–45\,°C$
Thermophils $> 45\,°C$

In tropical and subtropical sewage most bacteria are, as would be expected, mesophilic.

10

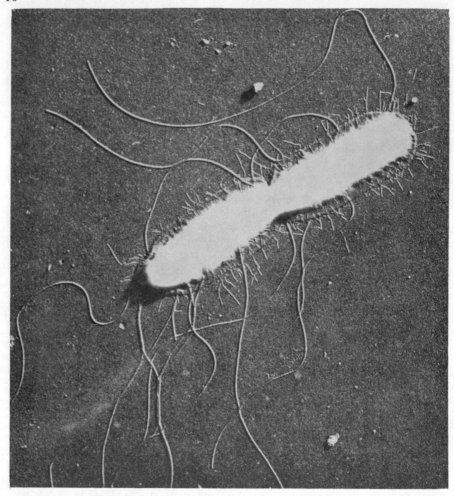

Figure 2.2 A dividing salmonella × 16000. The long hair-like appendages give bacteria the ability to move and the shorter ones offer some protection against predators [From J. P. Duguid and J. F. Wilkinson, in *Microbial Reaction to the Environment*, Cambridge University Press, London, 1961]

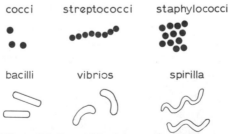

Figure 2.3 Bacterial shapes. Bacilli are the most common bacteria in sewage treatment. Typical sizes are: cocci, 1 mm diameter; bacilli, 1 × 2–5 mm; vibrios, 1 × 5 mm; spirillae, 1–2 × 5–20 mm

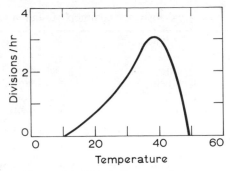

Figure 2.4 The influence of temperature on the rate of growth of a typical mesophil

Few bacteria can tolerate acid or alkaline conditions and most bacteria grow only in the near-to-neutral range of pH 5–9, with optimum growth usually occurring between pH 6 and 8.

Pathogenic bacteria

The important disease-causing bacteria commonly found in sewage are those that cause intestinal diseases, e.g. cholera, dysentery, typhoid and paratyphoid fever and diarrhoea. These organisms are highly infective and are responsible for many thousands of deaths in the tropics each year. Their spread is encour-

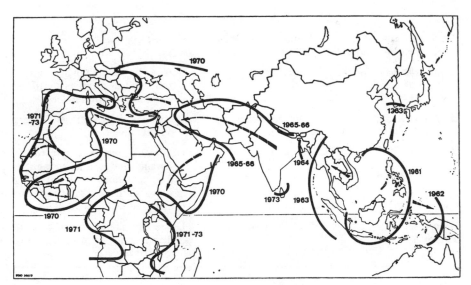

Figure 2.5 The global spread of cholera, 1961–1973 [From *WHO Weekly Epidemiological Record* **49**, 229 (1974)]

12

Table 2.1 Pathogenic organisms commonly found in sewage*

Organism	Disease	Remarks
Virus	Poliomyelitis Hepatitis	Exact mode of transmission not yet known. Found in effluents from biological sewage purification plants
Vibrio cholerae[+]	Cholera	Transmitted by sewage and polluted waters
Salmonella typhi[+]	Typhoid fever	Common in sewage and effluents in times of epidemics
Salmonella paratyphi[+]	Paratyphoid fever	Common in sewage and effluents in times of epidemics
Salmonella spp.[+]	Food poisoning	Common in sewage and effluents
Shigella spp.[+]	Bacillary dysentery	Polluted waters main source of infection
Bacillus anthracis[+]	Anthrax	Found in sewage. Spores resistant to treatment
Brucella spp.[+]	Brucellosis—Malta fever in man. Contagious abortion in sheep, goats and cattle	Normally transmitted by infected milk or by contact. Sewage also suspected
Mycobacterium tuberculosis[+]	Tuberculosis	Isolated from sewage and polluted streams. Possible mode of transmission. Care with sewage and sludge from sanatoria
Leptospira iceterohaemorrhagiae[+]	Leptospirosis (Weil's disease)	Carried by sewer rats
Entamoeba histolytica[+]	Dysentery	Spread by contaminated waters and sludge used as fertilizer. Common in hot climates
Schistosoma spp.	Schistosomiasis	Probably killed by efficient sewage purification
Taenia spp.	Tapeworms	Eggs very resistant, present in sewage sludge and sewage effluents. Danger to cattle on sewage-irrigated land or land manured with sludge
Ascaris spp. Enterobius spp.	Nematode worms	Danger to man from sewage effluents and dried sludge used as fertilizer

*From H. A. Hawkes, in *Microbial Aspects of Pollution*, Academic Press, London, 1971.
[+]These organisms are bacteria.

aged by the poor sanitary conditions and the general lack of hygiene which are so common in hot climates. A dramatic example of this is the spread of cholera through the tropics from southeast Asia in 1961 to central Africa a decade later (Figure 2.5).

Not all the pathogenic organisms in sewage are bacteria, although these are the most numerous. Pathogens commonly found in sewage are listed in Table 2.1.

Because it is rather difficult and very time-consuming to look for pathogenic organisms in a water or sewage sample, sanitary engineers look for a group of non-pathogenic bacteria which are easier to detect and which are always present in water which has been polluted by sewage or faeces. This group is the coliform group of bacteria.

Coliform bacteria

These bacteria are always present in very large numbers in faeces—the average adult excretes about 2 000 000 000 coliforms each day.[1] Their presence in water indicates that faecal pollution of the water has occurred and that the water may therefore contain pathogenic organisms. There are two major groups of coliforms: the faecal and the non-faecal coliforms. (Faecal coliforms are sometimes referred to as *Escherichia coli* type I (or just *E.coli*) and non-faecal coliforms as *Enterobacter* (or *Klebsiella* or *Aerobacter*) *aerogenes*. The coliform group as a whole is often called the *coli-aerogenes* group. It is just as well to steer clear of the troubled waters of bacterial nomenclature and simply refer to faecal and non-faecal coliforms.) The natural and exclusive habitat of faecal coliforms is the intestine of man and other warm-blooded animals. The presence of faecal coliforms in a water therefore indicates *beyond doubt* that faecal pollution has occurred. The position with non-faecal coliforms is not so well defined because they exist naturally unpolluted soils as well as in the intestine; their presence in water does not therefore *necessarily* imply faecal pollution. However, in practice the presence of non-faecal coliforms is taken as presumptive evidence of faecal pollution, particularly in hot climates where a significant proportion of faeces and sewage (the faecal origin of which cannot of course be held in doubt) may not contain any faecal coliforms.[2]

The numer of faecal coliforms in a sewage effluent is a reliable measure of its general bacteriological quality and in some countries standards have been set for the maximum permissible number of faecal coliforms in sewage effluent (Section 3.4). In Europe and North America standards are usually set for 'total coliforms' (i.e. faecal + non-faecal coliforms). This is not recommended in hot climates as the high ambient temperature commonly permit growth of the non-faecal coliforms, but not usually of faecal coliforms or pathogens; this has the result that the numbers of total coliforms are not reduced to the same degree as those of the faecal coliforms which are therefore better indicators of the possible presence of pathogens.

Transfer of drug resistance

Apart from the few enteropathogenic strains of faecal coliforms which cause fatal diarrhoea in newborn infants, coliform organisms are not pathogenic to man. However, they are not entirely harmless because they can transfer drug resistance to pathogenic organisms.[3] The resistance of a bacterium to a particular antimicrobial drug (an 'antibiotic') is the result of a genetic mutation and once the resistance gene is obtained, it is passed down from generation to generation. More importantly it can also be transferred from resistant to non-resistant cells. This is a very serious situation indeed because it means that there are fewer effective drugs available to medical practitioners for the control of an outbreak of a disease caused by drug-resistant strains of a particular pathogen. Such outbreaks are becoming increasingly more common and in hot climates particularly they represent possibly the gravest threat to public health witnessed this century. The pandemic of bacillary dysentery (shigellosis) which affected Central America in 1968–1971 was caused by a strain of the pathogenic bacterium *Shigella dysenteriae* which was resistant to chloramphenicol, streptomycin, sulphonamide and tetracycline (all drugs of choice for this disease); in January–October 1969 there were 112 000 cases of dysentery and 8200 deaths in Guatemala alone, where the main vehicle of transmission of the disease was contaminated water.[4,5]

Approximately 2 per cent of the coliforms from normal healthy people have drug resistance genes—this amounts to a discharge of 40 000 000 drug resistant coliforms per person per day. Because of the ease with which coliforms can accept and transfer drug resistance there is a need to reduce their numbers in the environment. This is most easily achieved by the proper design of sewage treatment facilities (see Section 7.10).

The bacteria of sewage treatment

The bacteria important in the aerobic treatment of sewage are rod-shaped, facultative and mesophilic. It is not necessary to describe them in detail; it is sufficient to know that they are excellent oxidizers of dead organic matter ('saprophytes') which grow extremely well in sewage. They are all capable of exuding a slimy flocculent layer which in some treatment units (e.g. activated sludge) is an important mechanism in the treatment process. Coliforms and other intestinal bacteria do not play any significant role in the sewage treatment processes; they are merely passengers in the system.

Bacterial growth

Bacteria can grow in an environment only if (1) there are sufficient nutrients available, (2) there is an absence of toxic compounds and (3) the environment itself is suitable. Bacteria require relatively large amounts of carbon, nitrogen, hydrogen and oxygen; smaller amounts of phosphorus, sulphur, potassium,

Figure 2.6 Mnemonic for the ten major elements essential to life: Carbon, Hydrogen, Oxygen, Phosphorus, Potassium (**K**), Nitrogen, Sulphur, Calcium, Iron (**Fe**) and Magnesium

calcium, iron and magnesium; and 'trace' quantities of several other elements (e.g. zinc, molybdenum); the ten major elements are easily remembered by the mnemonic shown in Figure 2.6. Environmental factors of major importance are: (1) neutral pH (about 6·5–8·5 pH units); (2) correct concentration of dissolved oxygen (zero for anaerobes; a minimum of 1–2 mg/l for aerobic growth); and (3) temperature (Figure 2.4).

Bacteria grow by subdivision into two daughter cells (Figure 2.2); this method of reproduction is called 'asexual binary fission'. In the laboratory some bacteria can complete the process of subdivision in as little as 10 minutes; in natural environments the process is much slower. Suppose a bacterium takes T minutes to subdivide into two: after nT minutes it will have produced a progeny of 2^n cells. This type of growth is *logarithmic* and is described by the equation:

$$N_t = N_0 e^{\mu t} \tag{2.1}$$

where N_t = number of cells at time t
N_0 = number of cells originally present at zero time
μ = specific growth rate; this term is more clearly defined by the differential form of equation 2.1, $dN/dt = \mu N$ (the units of μ are reciprocal time, e.g. d^{-1})
t = time.

If $N_t = 2 N_0$ then $t = T$, the time taken for the initial population to double in number, 'the mean generation time'). Putting these values into equation 2.1 we obtain:

$$2 N_0 = N_0 e^{\mu T}$$

which provides the following relationship between μ and T:

$$\mu = \frac{\ln 2}{T} \tag{2.2}$$

Batch culture curve

If we introduce a few bacteria into, say, a litre of soluble waste and if no further additions of waste are made, the bacteria will typically exhibit four distinct phases of growth (Figure 2.7). First is the *lag phase* during which time cell numbers do not increase; the bacteria are however internally active, manufacturing if necessary any intracellular catalysts ('enzymes') that they may require

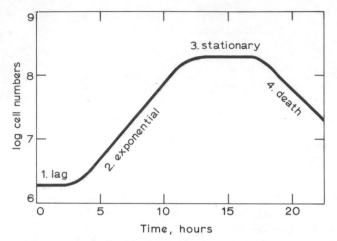

Figure 2.7 The bacterial batch culture curve (axis numbers
are illustrative only)

in order to be able to oxidize the waste. Next comes the *exponential phase*
during which logarithmic growth occurs; during this phase the bacteria lay
down food reserves within their cells which they may use when there is little or
no food left in their environment. The bacteria are now growing as fast as
they are able to in the waste; equation 2.1 is therefore rewritten as:

$$N = N_0 e^{\hat{\mu}t} \tag{2.3}$$

where $\hat{\mu}$ = maximum specific growth rate.

The exponential phase ceases, often abruptly, because either the supply
of an essential nutrient has been exhausted or there has been an accumulation
of toxic end-products of bio-oxidation (an example of the latter is the accumulation of acid which is an end-product of the oxidation of sugars; the pH falls
to a growth-inhibiting level). In the ensuing *stationary phase* the number of
new cells is approximately balanced by those that die, so that the population
cell density does not change. As the death rate exceeds the growth rate the
culture enters the *death phase* and the population density steadily declines.
During both the stationary and death phases there is a substantial proportion of
cells which neither die nor subdivide: they exist by utilizing the intracellular
food reserves laid down during exponential growth; this process is known as
endogenous respiration. When a cell has depleted its food reserves, it starts
to oxidize itself; this process, known as autolysis (= self-destruction), leads of
course to *death*.

Continuous culture

The biological treatment units commonly used in sewage treatment operate
continuously, 24 hours a day for 7 days a week, rather than as a batch process.

The growth rate of the bacteria in a continuous reactor is less than the maximum rate which occurs in batch culture; it depends on the sewage strength and the rate of autolysis:

$$\mu = \check{\mu} \left(\frac{L}{K_L + L} \right) - b \tag{2.4}$$

where L = BOD_5 of reactor contents, mg/l
$\quad K_L$ = 'saturation' constant (= value of L when $\mu = \mu/2$), mg/l
$\quad b$ = autolysis rate, d^{-1}.

Equations similar to equation 2.4 are commonly used in rational design procedures for activated sludge units.[6] A simplified model which incorporates the effect of autolysis is presented in Section 8.2 for the design of aerated lagoons.

2.3 VIRUSES

Viruses are peculiar microbes in that they do not directly use organic or inorganic compounds during growth; they reproduce by invading a host cell whose reproductive processes they redirect to manufacture more virus particles. They are extremely small (about 0·02–0·2 μm long) and when they are inactive they behave as stable chemical molecules and thus can remain infective for many years. The human diseases caused by viruses which are known to be, or may be, waterborne include poliomyelitis, smallpox and hepatitis.

2.4 ALGAE

Algae are mostly multicellular photosynthetic organisms, which are extremely varied in their shapes and sizes (Figure 2.8). Carbon dioxide is used as the source of carbon for the synthesis of new cells and oxygen is evolved from water by the classic mechanism of plant photosynthesis:

$$6\,CO_2 + 12H_2O \xrightarrow{\text{light}} \underset{\text{glucose}}{C_6H_{12}O_6} + 6\,H_2O + 6O_2$$

In darkness algae need oxygen for respiration and organic compounds for growth. Their growth (whether in darkness or light) is greatly stimulated by phosphates and nitrates which are usually present in sewage effluents. These salts cause nutrient-enrichment ('eutrophication') of a body of water and extensive algal growth then occurs—this is called an *algal bloom*. In the operation of waste stabilization ponds (Chapter 7) algal blooms are an essential part of the treatment process.

2.5 PROTOZOA

Protozoa are multicellular protists ancestral to the animal kingdom. There are three main groups of protozoa: amoebae, cilliates and flagellates. Amoebae and flagellates are not very important in sewage treatment, but the amoeba

18

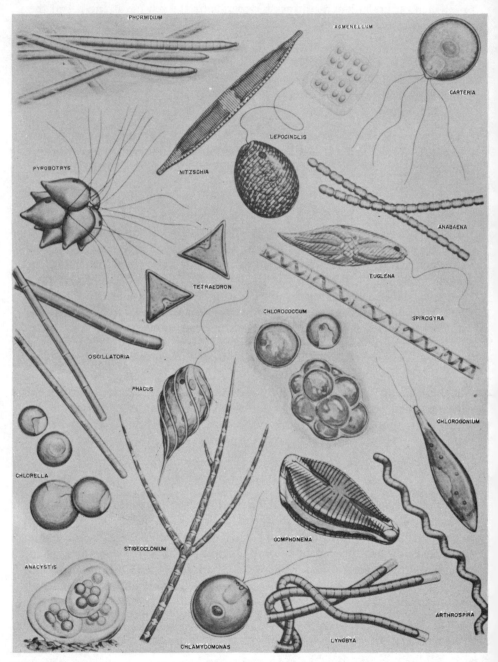

Figure 2.8 Algae commonly found in stabilization ponds and polluted waters [From C. M. Palmer, *Algae in Water Supplies*, US Public Health Service, Washington (Publication 657), 1962]

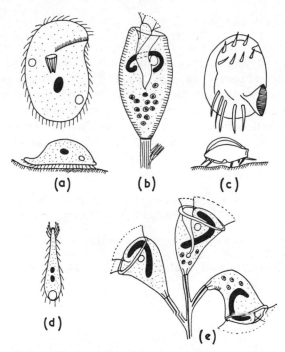

Figure 2.9 Five of the commonest ciliated protozoa found in sewage treatment works: (a) *Chilodonella uncinata*, (b) *Opercularia microdiscum*, (c) *Aspidisca costata*, (d) *Trachelophyllum pusillum* and (e) *Carchesium polypinum;* (a) and (c) are crawling ciliates, (b) and (e) stalked ciliates and (d) a free-swimming ciliate [From *Notes on Water Pollution No.* 43, HMSO, London, 1968]

Entamoeba histolytica is an important human pathogen causing amoebic dysentery (amoebiasis). The ciliates (Figure 2.9) are more important, being extremely common in sewage treatment works where they consume considerable numbers of bacteria; their numbers are about $10^3 - 10^4$ per ml. Controlled laboratory experiments have shown that they are responsible for a considerable proportion of the purification of sewage (Table 2.2).

Table 2.2 Results of laboratory experiments in which domestic sewage was treated in the absence and presence of ciliated protozoa*

Effluent property	Ciliates absent	Ciliates present
COD (mg/l)	198–254	124–142
Organic nitrogen (mg/l)	14–20	7–10
Suspended solids (mg/l)	86–118	26–34
Bacteria (millions/ml)	292–422	90–121

*From *Notes on Water Pollution No. 43*, HMSO, London, 1968.

2.6 INTESTINAL PARASITES

There is a large number of parasites which invade the human intestine and cause diseases of varying severity. Those which are waterborne include *Entamoeba histolytica* and *worms* of various kinds. The three main groups of parasitic worms are *nematodes* (roundworms), *cestodes* (tapeworms) and *trematodes* (flukes). The spread of these parasitic worms occurs largely through the improper or uncontrolled disposal of faeces—they therefore become increasingly more numerous with increasing lack of sanitation and personal hygiene.

2.7 REFERENCES

1. Geldreich, E. E., *Sanitary Significance of Faecal Coliforms in the Environment*, US Federal Water Pollution Control Administration, Washington, DC (Report WP-20-3, 1966).
2. Rao, N. U. *et al.*, *Environmental Health*, Nagpur, **10**, 21 (1968).
3. Grabow, W. O. K., *Water Research*, **8**, 1 (1974).
4. Gangarose, E. J. *et al.*, *Journal of Infectious Diseases*, **122**, 181 (1970).
5. Farran, W. E. and Edison, M., *Journal of Infectious Diseases*, **124**, 327 (1971).
6. Lawrence, A. W. and McCarty, P. L., *Journal of the Sanitary Engineering Division, American Society of Civil Engineers*, **96**, 757 (1970).

Further reading

D. D. Mara, *Bacteriology for Sanitary Engineers*, Churchill-Livingstone, Edinburgh, 1974.
R. E. McKinney, *Microbiology for Sanitary Engineers*, McGraw-Hill, New York, 1962.
J. R. Postgate, *Microbes and Man*, 2nd edition, Penguin, Harmandsworth, 1975.

3

Principles of Sewage Treatment

3.1 WASTEWATER MANAGEMENT

There are three constituent and interrelated aspects of wastewater management:

(1) Collection
(2) Treatment
(3) Re-use

Collection of domestic wastewater is best achieved by a full sewerage (water-carriage) system. Unfortunately this method is the most expensive and there are relatively few communities in hot climates which are able to afford it. A modern hygienic method of nightsoil collection is the only realistic alternative (Chapter 14). Treatment is required principally to destroy pathogenic agents in the sewage or nightsoil and to ensure that it is suitable for whatever re-use process is selected for it. The responsible re-use of nightsoil and sewage effluent in aquaculture and crop irrigation (Chapter 14) can make a significant contribution to a community's food supply and hence its general social development. The most striking example of the agricultural re-use of domestic wastes is in China: over 90% of the national nightsoil production is, after treatment, applied to the land and this represents *fully one-third* of all nutrients actually used by the crops.[1] In many of the tropical developing countries, which can ill afford to squander scarce resources, no attempt is made to re-use domestic wastes. The usual policy (insofar as there is one) is to follow that adopted by industrialized countries in temperate climates, namely the discharge of sewage effluent into rivers. This method of sewage disposal creates an entirely different emphasis in sewage treatment: a high degree of removal of organic matter (BOD) is required in order to prevent pollution of the receiving watercourse by oxygen depletion (Section 6.4). Industrialized countries may be able to afford to waste the nutrients in nightsoil and sewage effluent; tropical developing countries generally cannot.

Performance criteria for wastewater management systems[2]

The ideal system would satisfy all of the following criteria:

(1) *Health criteria.* Pathogenic organisms should not be spread either by

direct contact with the nightsoil or sewage or indirectly via soil, water or food. The treatment chosen should achieve a high degree of pathogen destruction.
(2) *Re-use criteria.* The treatment process should yield a safe product for re-use, preferably in aquaculture and agriculture.
(3) *Ecological criteria.* In those cases (and these should be considered exceptional) when the waste cannot be re-used, the discharge of effluent into a surface water should not exceed the self-purification capacity of the recipient water.
(4) *Nuisance criteria.* The degree of odour release must be below the nuisance threshold. No part of the system should become aesthetically offensive.
(5) *Cultural criteria.* The methods chosen for waste collection, treatment and re-use should be compatible with local habits and social (religious) practice.
(6) *Operational criteria.* The skills required for the routine operation and maintenance of the system components must be available locally or are such that they can be acquired with only minimum training.
(7) *Cost criteria.* Capital and running costs must not exceed the community's ability to pay. The financial return from re-use schemes is an important factor in this regard.

However, no one system completely satisfies all these demands. The problem becomes one of minimizing disadvantages.

3.2 SEWAGE TREATMENT

Sewage treatment is a combination of physical and biological processes; occasionally chemical processes are additionally employed. The common physical processes are screening, comminution and the removal of grit and organic SS by sedimentation (Sections 5.3 and 6.2). Biological processes involve the agency of bacteria and algae and constitute by far the most important methods of sewage treatment, particularly in hot climates; the fundamentals of biological oxidation are discussed in Section 3.3. Chemical processes are not now in common use, although considerable interest is being shown in the physico-chemical reclamation of drinking water from sewage effluent. However, owing to the cost of materials and the high degree of operator skill, they are unlikely to be used in developing countries in the foreseeable future, the only possible exception being the chlorination of sewage effluent (Section 11.3).

Siting of sewage treatment works

Sewerage networks are designed so that the flow of sewage is gravitational; pumping is avoided as much as possible in order to make the network independent of external power and so minimize costs and maintenance requirements. The location of sewage treatment works depends on the following general principles:

(1) Pumping should be avoided wherever possible.

(2) The site should be at least 500 m from the nearest house in order to avert any visual or odour nuisance; this is particularly important if anaerobic pretreatment ponds are to be used.

(3) The site should be of a suitable shape and have a gradient which permits gravitational flow from one process or pond to the next. If stabilization ponds are used the soil should preferably be impermeable.

(4) The site should be free from flooding.

(5) There should be adequate land available for future extensions.

Sewage flows

The flow and strength of a sewage varies throughout the day (Figure 3.1). The maximum or 'peak' flow arriving at the works is several times the mean flow and the hydraulic capacity of the works must be designed on the basis of the peak flow. The magnitude of the peak flow relative to the mean flow (the 'peaking factor') depends on the size of the contributing population: the larger the population, the lower the peaking factor since flow fluctuations are smoothed out during the time of travel in the sewer. The peaking factor M may be estimated from the formula:[3]

$$M = \frac{5}{P^{1/6}} \tag{3.1}$$

where P − contributing population in thousands.

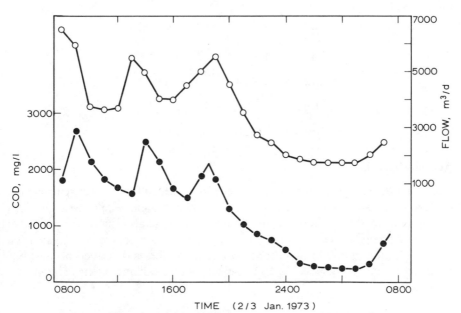

Figure 3.1 Diurnal variation in sewage flow (○) and strength (●) at Nakuru, Kenya

Alternatively the peaking factor may be chosen on the basis of sewer diameter:

Diameter	M
< 300 mm	2·5
300–600 mm	2·0
> 600 mm	1·5

Stormwater

In combined sewerage schemes the peaking factor can be very high indeed. It is usual to give full treatment to only three times the mean flow in dry weather (this flow is the true sewage flow and is called the 'dry weather flow', DWF). At conventional sewage works (Chapter 6) flows of 3–6 × DWF are given preliminary treatment and a short period of sedimentation in special 'storm' tanks reserved for this purpose before discharge to a river. Flows > 6 × DWF are usually only screened before discharge.

3.3 BIOLOGICAL OXIDATION

From the point of view of sewage treatment, bacteria can be considered as tiny automatic chemical reactors. They are considerably cheaper to operate than man-made reactors: they are self-adjusting and self-maintaining. Biological oxidation is the bacterial conversion of elements from organic forms to their most highly oxidized inorganic forms, a process known as *mineralization*; for example:

$$\text{Organic-C} + O_2 \xrightarrow{\text{bacteria}} CO_2$$

$$\text{Organic-H} + O_2 \xrightarrow{\text{bacteria}} H_2O$$

$$\text{Organic-N} + O_2 \xrightarrow{\text{bacteria}} NO_3^-$$

$$\text{Organic-S} + O_2 \xrightarrow{\text{bacteria}} SO_4^{2-}$$

$$\text{Organic-P} + O_2 \xrightarrow{\text{bacteria}} PO_4^{3-}$$

Aerobic oxidation

Bacteria oxidize wastes to provide themselves with sufficient energy to enable them to synthesize the complex molecules such as proteins and polysaccharides which are needed to build new cells. Thus bacterial metabolism has two component parts: *catabolism* (literally 'breaking down') for energy and *anabolism* (literally 'building up') for synthesis. Waste oxidation is described as 'aerobic' when molecular oxygen is used as the terminal oxidizing agent. The verbal 'equation'

$$\text{Wastes} + \text{Oxygen} \xrightarrow{\text{bacteria}} \text{Oxidized waste} + \text{New bacteria}$$

is instructive but oversimplified in that the anabolic and catabolic reactions are not distinguished; nor is there mention of autolysis which is an important form of catabolism. The following three equations describe these processes separately:

(1) *Catabolism*

$$\underset{\text{organic matter}}{C_xH_yO_zN} + O_2 \xrightarrow{\text{bacteria}} CO_2 + H_2O + NH_3 + \text{energy}$$

(2) *Anabolism*

$$C_xH_yO_zN + \text{energy} \xrightarrow{\text{bacteria}} \underset{\text{bacterial cells}}{C_5H_7NO_2}$$

(3) *Autolysis*

$$C_5H_7NO_2 + 5\,O_2 \xrightarrow{\text{bacteria}} 5\,CO_2 + NH_3 + 2\,H_2O + \text{energy}$$

As a general guide one-third of the available BOD is used in catabolic reactions and two-thirds in anabolic reactions (Figure 3.2). The equation for autolysis does not proceed to completion since approximately 20–25 per cent of the cell mass is resistant to aerobic degradation,

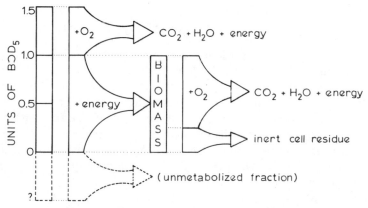

Figure 3.2 The catabolic, anabolic and autolytic reactions of aerobic biological oxidation. In a real (finite time) continuous biological reactor some of the organic matter in the influent escapes oxidation; in batch culture at infinite time the unmetabolized fraction is zero

Nutrients

Domestic sewage contains approximately the correct balance of nutrients required for bacterial growth, but some industrial wastes have insufficient nitrogen and phosphorus. The ratio BOD_5:N:P should be about 100:5:1.

Anaerobic digestion

Biological sludge treatment is commonly an anaerobic process, although

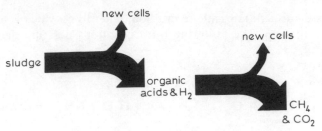

Figure 3.3 The acid-forming and methanogenic phases of
anaerobic digestion

aerobic digesters are sometimes used. Sewage solids are degraded in two stages
by two groups of anaerobic bacteria: first the organic compounds are oxidized
to fatty acids, mainly acetic acid, which are then converted to methane (Figure
3.3). For example the amino acid cysteine is degraded as follows:[4]

$$4C_3H_7O_2NS + 8H_2O \rightarrow 4CH_3COOH + 4CO_2 + 4NH_3 + 4H_2S + 8H$$
$$\text{cysteine} \qquad\qquad \text{acetic acid}$$

Methane-producing bacteria now convert the end-product of the acid forming
bacteria to methane:

$$4\,CH_3COOH + 8\,H \rightarrow 5\,CH_4 + 3\,CO_2 + 2\,H_2O$$

The acid-forming bacteria grow much more quickly than the methane producers, having doubling times of a few hours compared with a few days for
the latter group. It is important to ensure that there is a large healthy population
of methane producers since, without them, the sludge merely putrefies with
the attendant vile odours of decay; in contrast alkaline digestion with methane-
producing bacteria is an odourless process. The methane bacteria are sensitive
to acid conditions and the sludge liquor should have a minimum alkalinity
of 1000 mg/l as $CaCO_3$. The pH of the sludge is usually > 7; a value near 6
indicates imminent process failure.

Nitrification

Urea is the principal form in which the human body excretes excess nitrogen;
it is rapidly hydrolysed to ammonia. Nitrification is the bio-oxidation of
ammonia to nitrate. This conversion is a two-stage process which is accom-
plished by two groups of bacteria of the type that obtain their cell carbon from
carbon dioxide and their energy from the oxidation of inorganic compounds
(in this case, ammonia and nitrite). The bacteria are *Nitrosomonas* which
oxidizes ammonia to nitrite, and *Nitrobacter* which oxidizes nitrite to nitrate.
The formal equations are:[5]

Nitrosomonas

$$55NH_4^+ + 76O_2 + 5CO_2 \rightarrow C_5H_7NO_2 + 54NO_2^- + 52H_2O + 109H^+$$
$$\text{cells}$$

Nitrobacter

$$400NO_2^- + 195O_2 + 5CO_2 + NH_3 + 2H_2O \rightarrow C_5H_7NO_2 + 400NO_3^-$$

Thus about 3·33 g of molecular oxygen are required by *Nitrosomonas* to oxidize 1 g of NH_4-N and 1·11 g by *Nitrobacter* to oxidize 1 g of NO_2-N.

Nitrifying bacteria grow extremely slowly with doubling times of 1–2 d; in comparison the common aerobic sewage bacteria have typical doubling times of 0·25–1·5 h. They are generally active to a noticeable degree only in reactors with relatively long retention times (e.g. aerated lagoons, stabilization ponds and oxidation ditches) or when the concentration of organic compounds is low (e.g. in the bottom 0·5 m of a low-rate trickling filter). Nitrification is desirable if the effluent is to be used for irrigation because its end-product—nitrate— is a valuable plant nutrient; it is also desirable when the effluent is discharged into a surface water because ammonia can be toxic to fish and nitrification is an inexpensive method of ammonia removal. In stabilization ponds, nitrate acts as an algal nutrient, thus reinforcing the symbiosis between algae and bacteria that is the basis of BOD removal in ponds (Section 7.2).

3.4 SURFACE WATER POLLUTION

When a sewage effluent is discharged into a body of surface water (such as a lake, stream or river), not only are valuable nutrients wasted but also the receiving water may (and in practice usually does) become polluted. We may define a river as polluted when its quality has deteriorated to such a level that it is no longer suitable for its intended purpose. A realistic policy for establishing local quality standards for surface waters must be based on the local pattern of surface water usage. In hot climates the most important uses of surface waters are generally fishing, irrigation and domestic (including potable) water supply. It is much too expensive to attempt to maintain the quality of a river water at a level suitable for drinking (even though the standards may be considerably below the WHO international standards); it is often cheaper to maintain the river in state suitable for coarse fishing and develop alternative sources for domestic supplies.[6] In areas where surface waters are used for personal and clothes washing, particular attention must be paid to preventing the water becoming an active vehicle for the transmission of schistosomiasis.

For a river to maintain a healthy and economic population of fish there must be adequate dissolved oxygen and a negligible concentration of toxic compounds such as heavy metals and pesticides. Although very detailed stream standards do exist in some developing countries,[6] a simple but effective yard-stick of river water quality is the dissolved oxygen (DO) concentration (expressed as a percentage of the saturation concentration, or solubility, of oxygen in the river water at the river water temperature):[7]

Quality	% Saturation
Good	> 90
Fair	75–90

Quality	% Saturation
Doubtful	50–75
Badly polluted	< 50

The degree of faecal contamination can be assessed by estimating the number of faecal coliforms present by the agar dip-slide technique (Appendix 1).

Self-purification in a river

If an effluent is discharged into a river it exerts a demand on the oxygen resources of the river. This removal of DO for waste stabilization must be balanced by an addition of oxygen. The most important source of oxygen for reoxygenation of the river is the atmosphere: there is a mass transfer of oxygen from the atmosphere across the water surface to the bulk water below. The rate of this transfer is proportional to the oxygen deficit in the water (i.e. the difference between the saturation concentration and the actual DO concentration). Thus the DO removal that occurs below the point of discharge of an effluent actually stimulates an increased rate of supply of oxygen from the atmosphere. This competition between deoxygenation and reoxygenation results in a DO profile which typically shows a distinct sag some distance below the point of discharge (Figure 3.4). In order to prevent the river becoming offensive there must be an

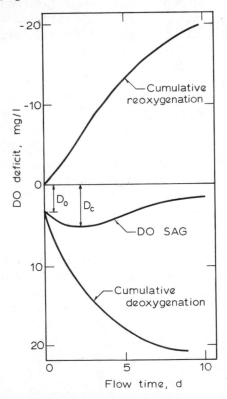

Figure 3.4 The oxygen sag curve is the sum of the deoxygenation and reoxygenation processes in the river (bacterial oxidation and surface re-aeration respectively). The waste is added to the river flow at zero time; the initial deficit D_0 decreases to a maximum or *critical* deficit D_c at a distance downstream equivalent to several days flow time

adequate DO reserve at all points along the river. Analysis of the oxygen sag curve[8] provides a convenient method of determining the degree of treatment that should be given to the effluent before it is discharged, so as to ensure that the lowest DO concentration that occurs is greater than the minimum required to maintain the river water quality at the desired level.

Effluent standards

It is administratively more convenient to enforce an effluent standard rather than a stream standard. The local regulatory agency should select quality standards for sewage effluent which ensure that the stream does not become unsuitable for its present use or intended purpose. The best known and most widely (and usually inappropriately) applied effluent standard is the so-called '20/30 Royal Commission standard' (20 mg BOD_5/l and 30 mg SS/l). The United Kingdom Royal Commission on Sewage Disposal of 1898–1915 was appointed to consider appropriate methods of sewage treatment and disposal *in Great Britain* (it is necessary to stress 'in Great Britain' because the Commissioners' recommendations were meant to apply only to this country, although they are often indiscriminantly applied to other countries in different climatic zones). The Commissioners classified British rivers on the basis of their 65 °F (18·3 °C) BOD_5 as follows:

Classification	BOD_5 (mg/l)
Very clean	>1
Clean	2
Fairly clean	3
Doubtful	5
Bad	>10

The Commissioners chose a 5 d, 65 °F BOD because British rivers do not have a flow time to the open sea > 5 d and the long-term average summer temperature in the UK is 65 °F. Thus a 5 d, 65 °F BOD was the maximum oxygen demand that a sewage effluent could exert in British rivers. (The standard BOD test is now conducted at 20 °C rather than 18·3 °C; thus since a 5 d, 18·3 °C BOD of 20 mg/l is equivalent to a 5 d, 20 °C BOD of about 23 mg/l, the present 20 °C standard is some 15 per cent stricter than the former 65 °F standard.)

If an effluent is discharged into a river, a mass balance of BOD at the point of discharge (Figure 3.5) yields:

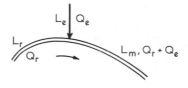

Figure 3.5

$$L_rQ_r + L_eQ_e = L_m(Q_r + Q_e) \qquad (3.2)$$

where $L = BOD_5$, mg/l $(= g/m^3)$

$Q =$ flow, m^3/d

r refers to the river upstream

e refers to the effluent

m refers to the river–effluent mixture just downstream of the point of discharge

Now if the effluent is diluted with 8 volumes of *clean* river water (i.e. if $Q_r/Q_e = 8$ and $L_r = 2$ mg/l), the maximum BOD_5 of the effluent to avoid nuisance in the river (i.e. for $L_m = 4$ mg/l), is given by equation 3.2 as:

$$\begin{aligned} L_e &= \frac{L_m(Q_r + Q_e) - L_rQ_r}{Q_e} \\ &= L_m[(Q_r/Q_e) + 1] - L_r(Q_r/Q_e) \\ &= 4(8 + 1) - (2 \times 8) \\ &= 20 \text{ mg/l} \end{aligned}$$

To this standard for BOD_5 the Commissioners added their standard of 30 mg/l for SS. Although the 20/30 standard is usually referred to as *the* Royal Commission standard, the Commissioners in fact laid down a standard for each of four ranges of dilution (Table 3.1). There is no evidence to suggest that the Royal Commission standards are directly applicable to climates other than those similar to that in Britain. The river classification used by the Commissioners is now considered too conservative even in Britain;[9] in hot climates the 'natural' BOD of a river can be high—for example, the unpolluted River Turkwell in the remote area of Northern Kenya has a BOD_5 of 20–45 mg/l.[10] In urban areas drainage and seepage from unsewered sections can pollute the river to such an extent that the effluent from a downstream sewage treatment works serving the sewered area can actually *improve* the river quality (Figure 3.6).

In most tropical developing countries effluent standards do not exist. Even

Table 3.1 UK Royal Commission standards for sewage effluents discharged into rivers

Available dilution (volumes of clean river water per unit volume of effluent)	Maximum permissible concentration (mg/l)	
	BOD_5	SS
> 500	—*	—*
300–500	—*	150
150–300	—*	60
8–150	20	30
< 8	<20[+]	<30[+]

*No standard recommended; theoretically infinite.
[+]Exact values to be decided on the basis of local circumstances.

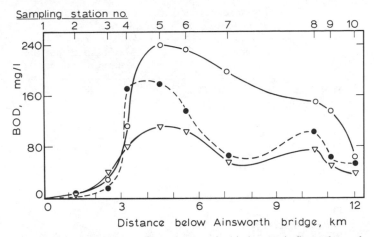

Figure 3.6 BOD profiles of the Nairobi river as it flows through Nairobi, Kenya. Waste flows from unsewered 'shantytown' developments below station 4 cause gross pollution of the river. Sewage effluent from the city's sewage treatment works are discharged below stations 6 and 9 and the resulting improvement in river quality is clearly evident. ○, low flow in river; ●, medium flow; ▽, high flow

so design engineers need to ensure that the effluent produced in their treatment works will not pollute the receiving watercourse. As a general guide the following *minimum* effluent standards should be adopted:

$$BOD_5 \quad < 25 \text{ mg/l}$$
$$FC \quad < 5000 \text{ cells/100 ml}$$
$$Algae \quad < 10^5 \text{ cells/ml}$$

No standard is suggested for SS since the natural SS concentration in tropical streams and rivers is usually high (turbidities are often > 100 FTU). The FC standard represents a very high degree of bacteriological purification, $> 99{\cdot}99$ per cent. The standard for algae is based on the finding[6] that the oxygen resources of the receiving watercourse would deteriorate if the concentration of algae in the watercourse were to exceed 1×10^5 cells/ml; in fact the final effluent from a properly designed series of stabilization ponds should always have an algal population $< 10^5$ cells/ml. The BOD_5 standard may seem high, but tropical streams with a BOD_5 of about this value do not generally give rise to nuisance. There would normally be some dilution so that the stream BOD_5 would be less. In the case of discharge into an ephemeral stream a more stringent standard may be required to prevent nuisance during the dry season.

There are occasions however when a more rigorous set of standards are called for: this is usually required where it is necessary to protect an area of prime biological importance, usually for the complementary benefits of conservation and tourism. One such example, for Lake Nakuru in Kenya, is given in Table 3.2.

Table 3.2 Quality standards for discharges of municipal sewage
effluent into Lake Nakuru, Kenya*

Parameter	Maximum permissible concentration (mg/l)
BOD_5 (excluding algal BOD_5)	50[+]
COD (excluding algal COD)	80[+]
SS	30
Ammonia (as N)	10[+]
Heavy metals (excluding Zn and Fe)	0·1
Zinc	0·3
Cyanide	0·05
Total phenols	0·1
Total organochlorine pesticides	0·001
Oil	No trace
Anionic detergents	0·5
Toxicity to fish	None[‡]

*Lake Nakuru is one of the series of alkaline soda lakes in the East African rift valley. It is the home of about two million lesser flamingoes and many other exotic birds; it has been described as 'the greatest ornithological spectacle in the world' (see L. Brown, 'The flamingoes of Lake Nakuru', *New Scientist*, 8 July 1971); it is now a national park, attracting many thousands of visitors each year. The Lake has no outlet and thus any toxic compounds entering it remain there for all time. The above standards, which are very strict indeed, were established with the sole purpose of protecting this unique heritage form any excessive accumulation of toxic material.

[+] These standards for BOD_5, COD and NH_3 are reduced to 50, 30 and 10 mg/l respectively for effluents discharged into any river draining into the lake (in fact waste stabilization pond effluent is discharged into the ephemeral Njoro river about 2 km from the lake).

[‡] An effluent is considered non-toxic if a 5 per cent dilution of it in lake water does not kill more than half the test population of *Tilapia grahami* (the only fish in the lake) within 2 d.

Discharge into coastal and estuarine waters

The sewage from coastal towns is discharged, usually with no treatment at all, into the sea via an outfall pipe. This pipe should be sufficiently long so that the sewage is always discharged below mean low water level. The actual location of the outfall depends largely on the pattern of the local tidal currents, but it should be chosen so as to:

(1) Ensure that there is no increased health risk to swimmers.
(2) Prevent the fouling of beaches with sewage solids of recognizable origin.
(3) Minimize damage to the marine ecosystem (particularly coral reefs).

The first two criteria are especially important if there is a large tourist industry. The third criterion is of long-term importance, particularly if the sewage contains an appreciable proportion of toxic industrial wastes. Although discharge to sea is the easy way of sewage disposal for coastal towns, sewage treatment and effluent re-use should always be considered as an alternative solution, especially since many coastal areas in hot climates are short of water. Moreover, we should try not to add to, but rather to reduce, the pollution of the sea which is a global resource of considerable economic value and ecological importance.

3.5 REFERENCES

1. Chao, K., *Agricultural Production in Communist China*, University of Wisconsin Press, 1970.
2. Winblad, U., *Evaluation of Waste Disposal Systems for Urban Low-income Communities in Africa: a research study*, Scan Plan Coordinator, Copenhagen, 1972.
3. Grifft, H. M., *Waterworks & Sewage*, **92**, 175 (1945).
4. *Notes on Water Pollution No. 64*, HMSO, London, 1974.
5. Wezernak, C. T. and Gannon, J. J., *Applied Microbiology*, **15**, 1211 (1967).
6. Pescod, M. B., *Investigation of Rational Effluent and Stream Standards for Tropical Countries*, Asian Institute of Technology, Bangkok, 1974.
7. Key, A., *Bulletin of the World Health Organization*, **14**, 845 (1956).
8. Nemerow, N. I., *Scientific Stream Pollution Analysis*, McGraw Hill, New York, 1974.
9. Klein, L., *River Pollution*, vol. 3, Butterworths, London, 1958.
10. Meadows, B. S., in *Proceedings of a Seminar on Sewage Treatment*, University of Nairobi, 1974.

Further reading

R. G. Feachem, M. G. McGarry and D. D. Mara, *Water, Wastes and Health in Hot Climates*, Wiley, London, 1976.

Water Pollution Control in Developing Countries (Technical Report Series No. 404), World Health Organization, Geneva, 1968.

Disposal of Community Wastewater (Technical Report Series No. 541), World Health Organization, Geneva, 1974.

4

BOD Removal Kinetics

4.1 FIRST ORDER KINETICS

The rate at which organic matter is oxidized by bacteria is a fundamental parameter in the rational design of biological waste treatment processes. It has been found that BOD removal often approximates first order kinetics; that is to say, the rate of BOD removal (= rate of oxidation of organic matter) at any time is proportional of the amount of BOD (= organic matter) present in the system at that time. Mathematically this type of reaction is written as:

$$\frac{dL}{dt} = -k_1 L \tag{4.1}$$

where L = amount of BOD remaining (= organic matter to be oxidized) at time t

k_1 = first order rate constant for BOD removal (units: reciprocal time, usually d^{-1}).

The differential coefficient dL/dt is the rate at which the organic matter is oxidized, and the minus sign indicates a decrease in the value of L with time. Equation 4.1 is the differential form of the first order equation for BOD removal; it can be integrated to:

$$L = L_0 e^{-k_1 t} \tag{4.2}$$

where L_0 = the value of L at $t = 0$.

L_0 is the amount of BOD in the system before oxidation occurs; it is therefore the ultimate BOD (Section 1.2). The amount of BOD removed or 'satisfied' (= organic matter oxidized) plus the amount of BOD remaining (= organic matter yet to be oxidized) at any time must obviously equal the ultimate BOD (= initial amount of organic matter):

$$y = L_0 - L \tag{4.3}$$

where y = the BOD satisfied at any time t.

Substitution into equation 4.2 yields:

$$y = L_0(1 - e^{-k_1 t}) \tag{4.4}$$

Generalized BOD curves (plots of equations 4.2 and 4.4) are shown in Figure 4.1 from which the relationship between y, L and L_0 is readily seen. Procedures

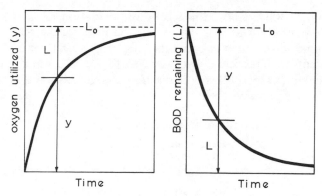

Figure 4.1 Generalized BOD curves

for analysing BOD data to determine obtain estimates of the values k_1 and L_0 are given in Appendix 2.

Equation 4.2 can be written in the form:

$$L = L_0 10^{-K_1 t}$$

where $K_1 = k_1/2.3$.

Because of the confusion that generally arises between K_1 and k_1, it is always best to give the base when quoting k_1 values, e.g. 0·23 (base e), 0·10 (base 10).

Ratio of BOD$_5$ to BOD$_u$

The ratio of $BOD_5/BOD_u (= y_5/L_0)$ is given by equation 4.4 as $(1 - e^{-5k_1})$. Under the conditions of the BOD test (Appendix 1) the value of k_1 for raw domestic sewage is typically 0·23 d^{-1} (base e; = 0·1 d^{-1}, base 10), so that the ratio BOD_5/BOD_u is about $\frac{2}{3}$.

Continuous flow processes

Equations 4.2 and 4.4 describe the bio-oxidation of a given quantity of organic matter to which no further addition is made. They represent conditions in a 'batch' oxidation process. But sewage treatment works operate with a *continuous* inflow of raw sewage and a continuous outflow of treated effluent.

Consider a mass balance of BOD across a continuously operated biological reactor: the quantity of organic matter entering the reactor per day must equal the quantity leaving the reactor per day plus that removed by bio-oxidation. Now if Q is the flow in m^3/d and L_i and L_e are the influent and effluent BOD respectively in mg/l (= g/m^3), then:

$$\left(\begin{array}{l} \text{quantity of BOD entering} \\ \text{the reactor, g/d} \end{array} \right) = L_i Q$$

$$\left(\begin{array}{l} \text{quantity of BOD leaving} \\ \text{the reactor, g/d} \end{array} \right) = L_e Q$$

The quantity of BOD removed in g/d by bio-oxidation *per unit volume* (m³) is given by equation 4.1 as $k_1 L$ where L is the BOD of the reactor contents. We will assume, for the moment, that the reactor is perfectly mixed so that the reactor contents are identical in every respect to the effluent from the reactor. Under this condition the BOD of the reactor contents is L_e. If V is the working volume of the reactor in m³ then:

$$\begin{pmatrix} \text{quantity of BOD removed} \\ \text{by bio-oxidation, g/d} \end{pmatrix} = k_1 L_e V$$

so that:

$$L_i Q = L_e Q + k_1 L V \tag{4.5}$$

Rearranging:

$$\frac{L_e}{L_i} = \frac{1}{1 + k_1 (V/Q)} \tag{4.6}$$

The ratio V/Q is the mean hydraulic retention time t^*, the average length of time a typical particle may be expected to remain in the reactor before being discharged in the effluent flow. Equation 4.6 can therefore be written as:

$$\frac{L_e}{L_i} = \frac{1}{1 + k_1 t^*} \tag{4.7}$$

This equation has found direct application in the design of waste stabilization ponds and aerated lagoons (sections 7.9 and 8.2).

Temperature

The rate constant k_1 is a gross measure of bacterial activity and, in common with almost all parameters describing a biological growth process, its value is strongly temperature dependent. Its variation with temperature is usually described by an Arrhenius equation of the form:

$$k_T = k_{20} \theta^{T-20} \tag{4.8}$$

where k_T and k_{20} are the values of k_1 at T °C and 20 °C respectively and θ an Arrhenius constant whose value is usually between 1·01 and 1·09. Typical θ values for some sewage treatment processes are:

Waste stabilization ponds	1·05 –1·09
Aerated lagoons	1·035
Trickling filters	1·040
Activated sludge processes	1·005–1·030

θ values are themselves a function of temperature, decreasing with increasing temperature, but they are normally sensibly constant over a 10 degC or 15 degC range. Thus a θ value for the temperature range 5–20 °C will not the same as that for the range 20–35 °C or even 15–30 °C. Caution must therefore be exercised in adopting θ values found in temperate climates.

4.2 HYDRAULIC FLOW REGIMES

The flow of sewage through a biological reactor can approximate either complete mixing or plug flow. These two flow patterns represent two extreme or ideal conditions. In practice the hydraulic regime lies between these extremes and is commonly described as 'dispersed flow'.

Complete mixing

The influent to this ideal reactor is completely and instantaneously mixed with the reactor contents which are, as a result of the intense mixing, uniform in composition throughout. The effluent is identical therefore in every respect to the reactor contents. The removal of BOD is described by equation 4.7.

Plug flow

The contents of this ideal reactor flow through the reactor in an orderly fashion which is characterized by the complete absence of longitudinal mixing. The concept of plug flow is readily grasped by imagining the sewage, on arrival at the reactor, to be placed in watertight 'packets' which then travel along the length of the reactor—as if on a conveyor belt—with no transfer of material from one packet to another, although there is a complete mixing within each packet. Since each packet receives no additional BOD and loses none to a neighbouring packet, the removal of BOD within each packet is essentially a batch process so that BOD removal in a plug flow reactor follows equation 4.2. It is however convenient to adopt the notation used in equation 4.7 and rewrite equation 4.2 as:

$$L_e = L_i e^{-k_1 t^*} \qquad (4.9)$$

Dispersed flow

It is impossible to build a plug flow reactor in which there is no mixing between packets; in practice some longitudinal mixing always occurs. The degree of inter-packet mixing that takes place is usually expressed in terms of a dimensionless *dispersion number* δ defined as:

$$\delta = \frac{D}{ul} \qquad (4.10)$$

where D = the coefficient of longitudinal dispersion, m²/h
  u = mean velocity of travel, m/h
  l = mean path length of a typical particle in the reactor, m.

38

When there is no longitudinal dispersion (i.e. in the case of ideal plug flow) $\delta = 0$ and when there is infinite dispersion (i.e. complete mixing) $\delta = \infty$.

In a dispersed flow reactor $(0 < \delta < \infty)$ in which bio-oxidation occurs as a first order reaction, the removal of BOD is described by the Wehner–Wilhelm equation:[1]

$$\frac{L_e}{L_i} = \frac{4a \exp(1/2\delta)}{(1 + a)^2 \exp(a/2\delta) - (1 - a)^2 \exp(-a/2\delta)} \qquad (4.11)$$

where $a = \sqrt{(1 + 4k_1 t^* \delta)}$.

Equation 4.11 reverts to equation 4.9 when $\delta = 0$ and to equation 4.7 when $\delta = \infty$. This rather complicated equation is not itself used, but rather the Thirumurthi[2] chart prepared from it (Figure 4.2). This chart shows that for any given combination of k_1 and t^* maximum BOD removal is achieved in an ideal plug flow reactor, and least removal in a completely mixed reactor of the same size. Expressed in another way this means that for any given value of k_1 and any desired degree of BOD removal the required retention time is a minimum in a plug flow reactor and a maximum in a completely mixed reactor. A plug flow reactor is therefore always smaller than a completely mixed reactor designed to achieve the same removal of BOD.

Dispersion numbers are determined by chemical tracer studies with either common salt or a fluorescent dye (e.g. rhodamine B).[3] Some typical results are shown in Figure 4.3.

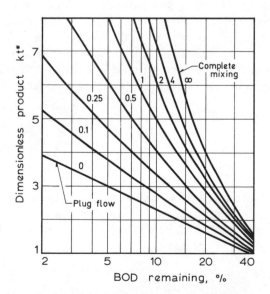

Figure 4.2 Thirumurthi chart for Wehner–Wilhelm equation. The numbers adjacent to each curve are the corresponding dispersion numbers

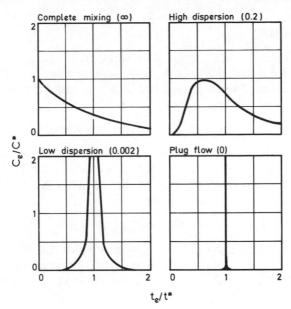

Figure 4.3 Typical tracer study results. A slug of dye is added to the influent and its concentration in the effluent C_e is determined at various corresponding times t_e. The results are plotted as the dimensionless numbers $C_e C^*$ and t_e/t^* where $C^* =$ the weight of dye added divided by the reactor volume and $t^* =$ mean hydraulic retention time (= reactor volume/flow rate).

In the completely mixed reactor the dye is instantaneously and uniformly distributed so that at zero time $C_e = C^*$; exponential wash out of the dye then follows. In the plug flow reactor all the dye appears at the effluent when $t_e = t^*$. Dispersed flow reactors behave in an intermediate fashion depending on the magnitude of their dispersion number; two examples are shown here, for $\delta = 0.2$ and $\delta = 0.002$

BOD$_u$ or BOD$_s$?

In equations 4.7, 4.9 and 4.11 which describe BOD removal in continuous flow reactors, the terms L_i and L_e may refer to either the ultimate or the 5 d BOD of the influent and effluent. In practice BOD$_5$ is most commonly used because it is more easily measured; k_1 is thus strictly interpreted as the first order rate constant for BOD$_5$ removal.

4.3 LIMITATIONS OF FIRST ORDER KINETICS

Equation 4.1 assumes that all the components of the waste are oxidized at the same rate and that the rate of oxidation remains constant with time. It is unlikely that all components of a waste so heterogeneous in nature as domestic sewage will be oxidized at the same rate, and it has been frequently observed in waste stabilization ponds, for example, that as the retention time increases the rate constant decreases.

Composite exponetial[4]

If it is assumed that different fractions of the waste are oxidized at different rates but that each rate constant does not decrease with time, the simple exponential term $e^{-k_1 t}$ in equation 4.4 can be replaced by a composite exponential so that the equation becomes:

$$y = L_0 [1 - f_1 \exp(-k_{f_1} t) - f_2 \exp(-k_{f_2} t) - \ldots - f_n \exp(-k_{f_n} t)] \qquad (4.12)$$

where $f_1, f_2, \ldots, f_n$ are the fractions of the waste oxidized at rates $k_{f_1}, k_{f_2}, \ldots, k_{f_n}$. For example an English sewage was found[4] to be oxidized as if it were a mixture of components of which 40 per cent were oxidized at a rate of $0\cdot8$ d^{-1} 40 per cent at $0\cdot08$ d^{-1} and 20 per cent at $0\cdot008$ d^{-1}:

$$y = L_0 (1 - 0\cdot4e^{-0\cdot8t} - 0\cdot4e^{-0\cdot08t} - 0\cdot2e^{-0\cdot008t})$$

The effluent from the sewage treatment works at which this particular sewage was treated, was oxidized as if it were composed of a 40 per cent fraction oxidizable at a rate of $0\cdot08$ d^{-1} and a 60 per cent fraction oxidizable at $0\cdot008$ d^{-1}:

$$y = L_0 (1 - 0\cdot4e^{-0\cdot08t} - 0\cdot6e^{-0\cdot008t})$$

These (typical) results show that the most rapidly oxidizable fraction was totally eliminated during treatment and, as a result, the effluent had a higher proportion of material that could be oxidized only very slowly.

Retarded exponential[4]

If, on the other hand, it is assumed that all the components in the waste are oxidized at the same rate but that the rate of oxidation decreases with time, then equation 4.7 is written as:

$$\frac{dL}{dt} = -\frac{k_1}{1 + \alpha t} L \qquad (4.13)$$

where $\alpha =$ a coefficient of retardation, d^{-1}.
k_1 is now defined as the rate constant at zero time. In its integrated form this equation is written as:

$$y = L_0 [1 - (1 + \alpha t)^{-k_1/\alpha}] \qquad (4.14)$$

The application of equation 4.13 to the design of waste stabilization ponds has not received much attention, yet it should be of considerable interest to know field values of α.

Second order kinetics[5,6]

Models of BOD removal based on second order kinetics have been found to describe the operation of certain biological treatment processes better than those based on first order kinetics. Equation 4.1 is replaced by:

$$\frac{dL}{dt} = -k_2 L^2 \tag{4.15}$$

where k_2 = second order rate constant for BOD removal, $(mg/l)^{-1} d^{-1}$
Equation 4.15 can be integrated to:

$$\frac{L}{L_0} = \frac{1}{1 + k_2 L_0 t} \tag{4.16}$$

Procedures for analysing BOD data to determine the values of k_2 and L_0 are given in Appendix 2. For a plug flow reactor equation 4.16 would be written as:

$$\frac{L_e}{L_i} = \frac{1}{1 + k_2 L_i t^*} \tag{4.17}$$

For a completely mixed reactor a mass balance of BOD (similar to that in equation 4.5) yields:

$$L_i Q = L_e Q + k_2 L_e^2 V \tag{4.18}$$

which is a quadratic equation in L_e. Substituting t^* for V/Q and solving for L_e gives:

$$L_e = \frac{-1 + \sqrt{(1 + 4k_2 L_i t^*)}}{2k_2 t^*} \tag{4.19}$$

There is no analytical solution for dispersed flow reactors; graphical solutions are however available.[3]

4.4 REFERENCES

1. Wehner, J. F. and Wilhelm, R. H., *Chemical Engineering Science*, **6**, 89 (1956).
2. Thirumurthi, D., *Journal of the Sanitary Engineering Division, American Society of Civil Engineers*, **95**, 11 (1969).
3. Levenspiel, O., *Chemical Reaction Engineering*, Wiley Eastern, New Delhi, 1967.
4. Gameson, A. L. H. and Wheatland, A. B., *Journal and Proceedings of the Institute of Sewage Purification*, (2), 106 (1958).
5. Young, J. C. and Clarke, J. W., *Journal of the Sanitary Engineering Division, American Society of Civil Engineers*, **91**, 43 (1965).
6. Tucek, F. and Chudoba, J., *Water Research*, **3**, 559 (1968).

4.5 WORKED EXAMPLES

1. *The BOD_5 of a waste has been measured as 600 mg/l. If $k_1 = 0.23 day^{-1}$ (base e) what is the BOD_u of the waste? What proportion of the BOD_u would remain unoxidized after 20 d?*
From equation 4.4:

$$y_5 = L_0(1 - e^{-k_1 \cdot 5})$$
$$\therefore L_0 = y_5(1 - e^{-k_1 \cdot 5})^{-1}$$
$$= 600(1 - e^{-1 \cdot 15})^{-1}$$
$$= 800 \text{ mg/l}$$

From equation 4.2:

$$\frac{L_{20}}{L_0} = e^{-k_1 \cdot 20}$$

$$= e^{-(0 \cdot 23)(20)}$$

$$= 0 \cdot 01$$

Thus 99 per cent of the waste has been oxidized in 20 d. BOD_{20} is therefore often taken as an approximation for BOD_u.

2. *Show that the ratio of the $2\frac{1}{2}$-day, 35 °C BOD to the 5-day, 20 °C BOD is approximately unity. Take θ as 1·05.*
From equation 4.4:

$$y_{2 \cdot 5} = L_0[1 - \exp(-2 \cdot 5 k_{35})]$$

and

$$y_5 = L_0[1 - \exp(-5 k_{20})]$$

But from equation 4.8:

$$k_{35} = k_{20}(1 \cdot 05)^{35-20}$$

$$= 2 \cdot 08 k_{20}$$

Substituting for k_{35} in the expression for $y_{2 \cdot 5}$:

$$y_{2 \cdot 5} = L_0[1 - \exp(-2 \cdot 5 \times 2 \cdot 08 k_{20})]$$

$$= L_0[1 - \exp(-5 \cdot 2 k_{20})]$$

$$\doteqdot y_5$$

Thus the BOD_5 of a waste or effluent can be obtained in $2\frac{1}{2}$ d if the test temperature is 35 °C rather than 20 °C—see H. R. Tool, *Water & Sewage Works*, **114**, 211 (1967).

3. *Show that in a biological reactor in which BOD removal follows second order kinetics, more BOD is removed under plug flow conditions that when the reactor contents are completely mixed.*
For second order kinetics the removal of BOD in a plug flow reactor is described by equation 4.17:

$$\left(\frac{L_e}{L_i}\right)_{\substack{\text{plug} \\ \text{flow}}} = \frac{1}{1 + k_2 L_i t^*}$$

For a completely mixed reactor equation 4.18 states:

$$L_i Q = L_e Q + k_2 L_e^2 V$$

Divide each term by $L_e Q$:

$$\frac{L_i}{L_e} = 1 + k_2 L_e (V/Q)$$

$$= 1 + k_2 L_e t^*$$

Take the reciprocal of each side:

$$\left(\frac{L_e}{L_i}\right)_{\substack{\text{complete} \\ \text{mixing}}} = \frac{1}{1 + k_2 L_e t^*}$$

(This is an alternative expression for equation 4.19.)

Now:
$$L_i > L_e$$

$$\therefore\ 1 + k_2 L_i t^* > 1 + k_2 L_e t^*$$

$$\therefore\ \frac{1}{1 + k_2 L_i t^*} < \frac{1}{1 + k_2 L_e t^*}$$

$$\therefore\ \left(\frac{L_e}{L_i}\right)_{\substack{\text{plug} \\ \text{flow}}} < \left(\frac{L_e}{L_i}\right)_{\substack{\text{complete} \\ \text{mixing}}}$$

The ratio (L_e/L_i) is the fractional BOD *remaining*. The BOD *removed* is $[1 - (L_e/L_i)]$.

$$\therefore\ \left(1 - \frac{L_e}{L_i}\right)_{\substack{\text{plug} \\ \text{flow}}} > \left(1 - \frac{L_e}{L_i}\right)_{\substack{\text{complete} \\ \text{mixing}}}$$

5

Preliminary Treatment

5.1 PURPOSE

The first stage of sewage treatment is usually the removal of large floating objects (such as rags, maize cobs, pieces of wood) and heavy mineral particles (sand and grit). This is done in order to protect from damage the equipment used in the subsequent stages of treatment (for example the floating aerators in aerated lagoons or any pumps which may be used). This preliminary treatment comprises *screening* and *grit removal*. A common alternative to screening is *comminution*.

At small works, particularly waste stablization ponds treating flows < 1000 m³/d, there is often no preliminary treatment, or at most only screening. This is commonly satisfactory, provided that the increased quantity of scum that forms on the surface of the first pond is, as all scum should be, regularly removed. At larger pond installations it is almost always advisable to include screening, or comminution, and grit removal. The operational advantages afforded by these processes increase directly with the volume of waste being treated.

5.2 SCREENING

Coarse solids are removed by a series of closely spaced mild steel bars placed across the flow. The velocity through the screen should be > 0.3 m/s in order to prevent the deposition of grit but < 1 m/s so that the solids already trapped on the screen ('screenings') are not dislodged. The spacing between the bars is usually 20–40 mm and the bars are commonly of rectangular cross-section, typically 10 mm × 50 mm. At small works screens are raked by hand and in order to facilitate this the screens are inclined, commonly at 60° to the horizontal (Figure 5.1). The submerged area of hand-raked screens is calculated on the empiricial basis of 0·15–0·20 m² per 1000 population; this assumes that the screens are raked at least twice each day.

For flows > 200 m³/d mechanically raked screens (Figure 5.2) are preferred since they can be cleaned more frequently (every 10–30 minutes) and are therefore considerably smaller than the corresponding hand-raked screen. The channel dimensions required for a mechanically raked screen are calculated as follows:

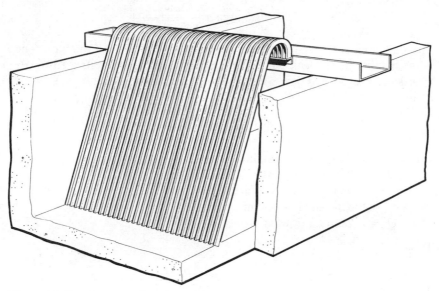

Figure 5.1 Simple manually raked screen. Flow is from left to right [Courtesy of Ames Crosta Ltd]

Figure 5.2 Mechanically raked screen, 'Shallomatic' [Courtesy of Ames Crosta Ltd]

$$\text{flow area} = \frac{\text{flow}}{\text{velocity}}$$

The flow area is the channel area corrected for the area of the bars. The flow is the maximum or 'peak' flow. The velocity is generally restricted to 0·6 m³/s in order to prevent grit deposition and dislodgement of screenings. The equation is therefore:

$$WD\left(\frac{s}{b+s}\right) = \frac{\hat{Q}}{0\cdot6} \tag{5.1}$$

where W = the channel width, m
  D = channel depth at maximum flow, m
  s = spacing between adjacent bars, mm
  b = the bar thickness, mm
  $\hat{Q}$ = maximum flow, m³/s.

Thus:

$$W = \frac{\hat{Q}}{0\cdot6D}\left(\frac{b+s}{s}\right) \tag{5.2}$$

A standby hand-raked screen should be provided for use when the mechanical screen is out of action. This emergency screen is normally the same size as the mechanical screen and it will therefore require raking at frequent intervals when in use.

Disposal of screenings

Screenings are particularly obnoxious in both appearance and content and should be disposed of as soon as possible. At small works this is readily achieved by burial, a small area of land being set aside for this purpose. At larger works screenings are either dewatered in a hydraulic press and then incinerated or macerated in a disintegrator and then returned to the sewage flow below the screens. Handling of screenings is a most unpleasant job and at larger works comminution is often preferred to screening for this reason alone.

The quantity of screenings that are removed varies considerably but, for for 10 mm bars at 20 mm spacings, an approximate figure is 0·01–0·03 m³/d per 1000 population. In South Africa screenings are removed at a rate of about 0·05 m³/1000 m³ of sewage treated.[1]

5.3 GRIT REMOVAL

Grit (also called 'detritus') is the heavy inorganic fraction of sewage solids. It includes road grit, sand, eggshells, ashes, charcoal, glass and pieces of metal; it may also contain some heavy organic matter such as seeds, coffee grounds, etc. Grit has an average relative density of about 2·5 and thus has a much higher settling velocity than organic sewage solids (about 30 mm/s as against 3 mm/s).

It is this difference in sedimentation rates that is exploited in grit removal plants where, for ease of handling and disposal, the organic fraction must be kept to a minimum (< 15 per cent). There are two basic types of grit removal plant: constant velocity grit channels and the various proprietary tanks or traps available commercially.

Constant velocity grit channels

If the velocity of flow of the wastewater is about 0·3 m/s, grit particles settle out but organic solids do not. The problem is to maintain the velocity constant at this value *for all rates of flow*. The best solution[2] is to locate a standing wave (venturi) flume immediately downstream of a grit channel of parabolic cross-section. This solution depends on the following two points:

(1) Provided that it is not 'drowned', a venturi flume produces an upstream depth that is independent of conditions downstream and which is controlled only by the magnitude of the flow
(2) If the geometry of the grit channel is such that its cross-sectional area is proportional to the flow, then the velocity of flow through the channel will be constant at all flows (if v = velocity, q = flow and a = cross-sectional area, then $v = q/a$; but if a is proportional to q, i.e. $a = kq$, then $v = q/kq$ = a constant).

It is now shown that to comply with (2) the channel should be of parabolic section. The flow q through a venturi flume is given by:

$$q = kbh^{3/2} \tag{5.3}$$

where k = a constant
b = throat width
h = upstream depth.
Differentiating:

$$dq = \tfrac{3}{2} kbh^{1/2}\, dh \tag{5.4}$$

The flow dq through a cross-sectional element of the channel (Figure 5.3) is given by:

$$dq = V x dh \tag{5.5}$$

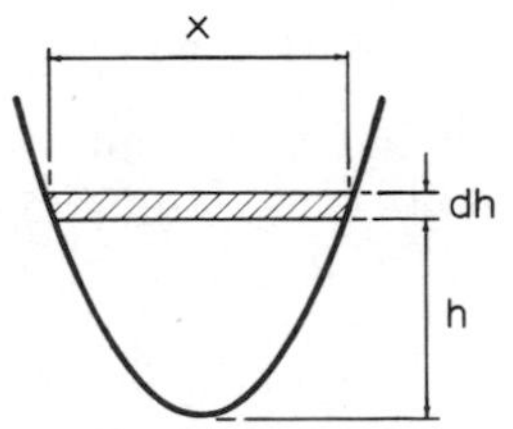

Figure 5.3

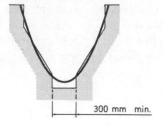

Figure 5.4 Trapezoidal approximation to parabolic section

300 mm min.

where V is the velocity of flow and $x\mathrm{d}h$ the area of the element. Equating equations 5.4 and 5.5 and rearranging gives:

$$x = \left(\frac{3kb}{2V}\right)h^{1/2} \tag{5.6}$$

This is the equation of a parabola. In practice, for ease of construction, a trapezoidal cross-section is used (Figure 5.4).

If $V = 0\cdot3$ m/s and X and H are the channel dimensions (m) at maximum flow $\hat{Q}$ (m^3/s), then equations 5.3 and 5.6 can be rewritten as:

$$\hat{Q} = kbH^{3/2} \tag{5.7}$$

and

$$X = 5\,kbH^{1/2} \tag{5.8}$$

Dividing equation 5.8 by equation 5.7 and rearranging gives:

$$x = \frac{5Q}{H} \tag{5.9}$$

Thus the top width of the channel is simply determined from the maximum flow and the corresponding depth. In practice however at least two channels are provided so that one may be closed for manual grit removal. The channel length is determined by the settling velocity of the grit particles:

$$\text{length of channel} = \frac{\text{channel depth} \times \text{velocity of flow}}{\text{settling velocity of grit particles}}$$

Grit particles typically settle through sewage at about $0\cdot03$ m/s, so that when the velocity of flow is controlled to $0\cdot3$ m/s:

$$\text{length of channel} = 10 \times \text{maximum depth of flow}$$

In practice, to allow for inlet turbulence and variations in settling velocity, the channel length is taken as $20 \times$ maximum depth of flow.

Proprietary grit separators

For flows > 5000 m^3/d proprietary grit separators are often more economical

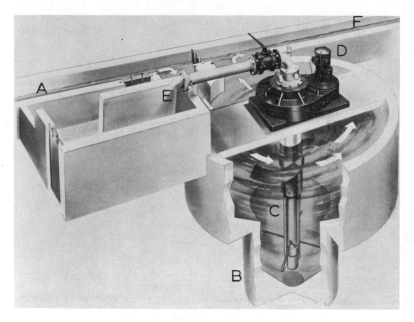

Figure 5.5 'Pista' grit trap. A, influent channel; B, grit trap; C, paddle stirrer and air lift pump with support legs; D, geared motor; E, grit discharge pipe; F, effluent channel [Courtesy of Jones & Attwood Ltd]

than several long constant-velocity grit channels. The most common type is a settling tank of sufficiently short retention time to allow the grit, but not the organic solids, to settle out. There are several models available. One of the simplest is the Jones & Attwood[3] 'Pista' grit trap which has the advantages that no moving parts come in contact with the grit and that the grit is automatically cleaned before discharge (Figure 5.5).

Grit disposal

The quantity of grit collected may be as high as 0.17 m^3/1000 m^3 of sewage,[1] although the average figure is 0.05–0.10 m^3/1000 m^3. The grit is either used for landfill or disposed of by burial.

5.4 COMMINUTION

A comminutor is a self-cleansing shredding machine which cuts up sewage solids as they pass or are pulled through a fine screen which forms the outer peripherly of the machine (Figure 5.6). It consists of a hollow cast-iron drum which is continuously rotated about its vertical axis by an electric motor through a reduction gearbox. The drum is in fact a screen with 6–8 mm horizontal slots on which cutter bars and a large number of projecting cutting

50

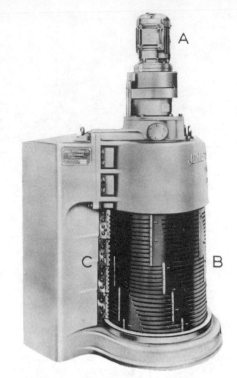

Figure 5.6 Comminutor. A, drive motor;
B, rotating drum. For detail C see Figure 5.7
[Courtesy of Jones & Attwood Ltd]

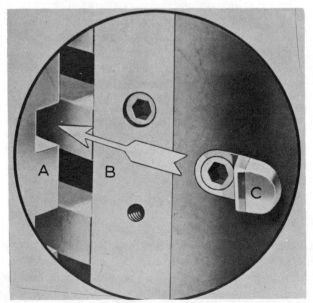

Figure 5.7 Relative positions of comb (A), cutting bar
(B) and cutting tooth (C) [Courtesy of Jones & Attwood Ltd]

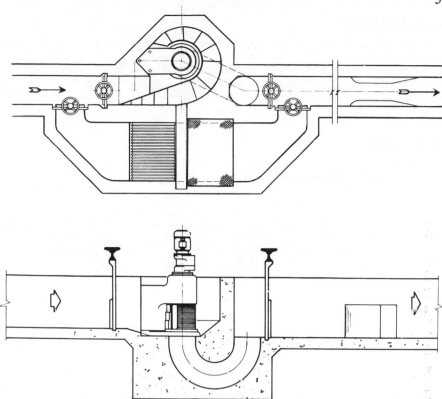

Figure 5.8 Typical single comminutor installation showing bypass channel with screen and inverted siphon below [Courtesy of Jones & Attwood Ltd]

teeth are fixed; the bars and teeth engage with stationary steel combs (Figure 5.7). The solids which are held against the outside of the moving drum by the sewage flow are carried by the drum to the stationary combs where they are shredded by the combined action of the cutter bars and cutting teeth. Comminutors are installed in special concrete chambers which are spiral in plan. Immediately below the machine there is an inverted siphon which connects the upstream and downstream channels (Figure 5.8). Routine maintenance is limited to lubrication and the replacement of cutting teeth and combs. A bypass channel which incorporates a manually raked screen should be provided for use during periods of maintenance or power failure.

Comminutors avoid the problems associated with the handling and disposal of screenings and for this reason they are popular with plant operators. However, shredded rags often re-form into lengths of string which 'ball up' and thus cause blockages in downstream pipes and pumps; wastes from textile factories are particularly troublesome in this respect. Grit can damage the cutting teeth and to avoid this occurring comminutors should be located downstream of grit channels or traps.

5.5 REFERENCES

1. Meiring, P. C. G. *et al.*, *CSIR Special Report WAT 34*, Council for Scientific and Industrial Research, Pretoria, 1968.
2. Townend, C. B., *Journal and Proceedings of the Institute of Sewage Purification*, (2), 58 (1937).
3. Jones & Attwood Ltd., Stourbridge, Worcestershire, U. K. (Publication No. 280 for Pista grit trap; No. 275A for comminutors).

Further reading

Manuals of British Practice in Water Pollution Control—1. Preliminary Processes, Institute of Water Pollution Control, Maidstone, Kent, 1972.

6

Conventional Treatment

6.1 INTRODUCTION

Conventional treatment is the term used to describe the standard method of sewage treatment in temperate climates. It comprises four stages of treatment (Figure 6.1):

(1) Preliminary treatment (as described in Chapter 5).
(2) Primary or physical treatment (sedimentation).
(3) Secondary or biological treatment (biofiltration or activated sludge).
(4) Sludge treatment (anaerobic digestion of the sludge produced in stages (2) and (3)).

Even though conventional treatment was developed in temperate climates *for* temperate climates, there are many existing conventional works in hot climates; and, although many of these were built before the advent of waste stabilization ponds, aerated lagoons and oxidation ditches (which are now

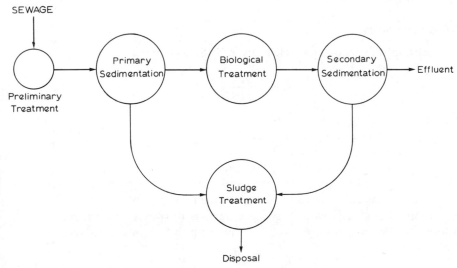

Figure 6.1 Basic flow diagram for conventional sewage treatment

generally the preferred methods in hot climates), there are a few instances of recent installations, for example the activated sludge installation at Kitwe, Zambia, capable of treating 58 000 m³/d.[1] Cost generally militates against conventional works (see Figure 7.5). Although in some instances economic analysis has shown that for large populations conventional treatment is cheapest, the necessary equipment has usually to be imported. A developing country with limited reserves of foreign exchange is likely to have other projects of equally high social priority which may have better claims for an allocation of foreign currency. The operational difficulties which beset conventional works in hot climates are discussed in Section 6.7. Nevertheless an understanding of conventional treatment is useful to engineers working in hot climates since existing conventional works have to be maintained and often altered; moreover, it has formed the basis for the development of design criteria for aerated lagoons, oxidation ditches and high-rate biofiltration with plastic media.

6.2 SEDIMENTATION

Sedimentation is the gravitational separation of a suspension into its component solid and liquid phases. In the primary sedimentation of sewage there are two aims: to produce high degrees of both clarification and thickening. Clarification is the removal of solids from the liquid phase and thickening the removal of liquid from the solid or sludge phase. A high degree of clarification is required to reduce the load on the secondary (biological) treatment plant and a high degree of thickening is desirable so that sludge handling and treatment (which usually accounts for 30 per cent of the total cost of conventional treatment[2]) is minimized.

Design principles

The settling velocity of a particle is given by Stokes law:

$$U_s = \frac{g\,(\rho_s - \rho)d^2}{18\,\mu} \qquad (6.1)$$

where U_s = settling velocity, m/s
$\quad g$ = acceleration due to gravity, m²/s
$\quad \rho$ = density of particles, kg/m³
$\quad \rho_s$ = density of suspending fluid, kg/m³
$\quad d$ = characteristic linear dimension of the particle (particle 'diameter'), m
$\quad \mu$ = molecular viscosity of suspending fluid, Ns/m².
Equation 6.1 is strictly valid only if the particle Reynolds number, R_e, is < 0.2, although little error is introduced for values up to 5. R_e is defined as:

$$R_e = \frac{\rho U_s d}{\mu} \qquad (6.2)$$

A particle can be assumed to be retained in a continuous flow sedimentation tank if its settling velocity is high enough to permit it to settle through the full tank depth within one retention time. Under this condition the limiting minimum settling velocity is:

$$U_s = D/t^* \qquad (6.3)$$

where D = tank depth, m
   t^* = mean hydraulic retention time, s.
The retention time in a continuous flow tank is given by:

$$t^* = AD/Q \qquad (6.4)$$

where AD = tank volume, m^3
   A = surface area of tank, m^2
   Q = flow through tank, m^3/s.
Therefore from equations 6.3 and 6.4:

$$U_s = Q/A \qquad (6.5)$$

The ratio Q/A is the flow per unit surface area and is termed the *overflow rate* (its units are those of velocity since m^3/m^{2}s = m/s). The traditional method of design is an empirical procedure based on this concept of hydraulic surface loading, incorporating values of overflow rate which experience has shown to be satisfactory. Thus for a typical design value of 30 m^3/m^2 d (which represents settlement through a tank depth of 2·5 m in 2 h) the smallest size of particle that will be removed in this period can be calculated from equation 6.1 (assuming ρ_s 1200 kg/m^3 and $\mu = 1·01 \times 10^{-3}$ N s/m^2 at 20 °C) as 57 μm. This figure is very much an approximation since Stokes law is strictly applicable only to particles in dilute suspension (< 1 per cent by volume); moreover, flocculation may occur and thus finer particles will also be removed. Other important effects which are ignored in the traditional design procedure are the solids concentration in the influent and in the sludge layer, the rate of sludge removal and the resuspension of solids from the sludge by turbulent shear at the sludge–liquid interface.[3] (The effects of the solids balance and the sludge flow are illustrated in the design example in Section 8.7. The effect of turbulent shear can be allowed for by using the design method of Clements.[4,5])

Tank geometry

Sedimentation tanks for sewage are now most commonly *radial flow tanks*. They are circular in plan, with small floor slopes ($2\frac{1}{2}$°–$7\frac{1}{2}$°). The sewage enters centrally, passes through an inlet baffle (to minimize turbulence) and then flows outwards and upwards to the overflow weir. Mechanical scrapers are provided for sludge collection and surface scum is removed by skimmer arms (Figure 6.2). The scraper arms rotate slowly (1–3 rev/h) and move the sludge towards a central sump from where it is periodically removed.

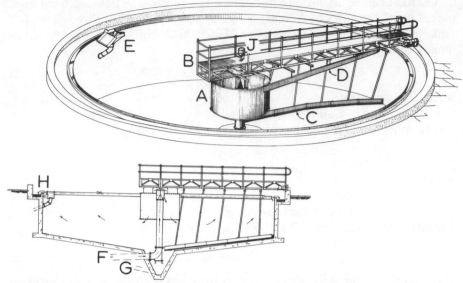

Figure 6.2 Perspective view and true elevation of circular radial flow sedimentation tank. A, inlet baffle; B, access bridge; C, sludge scrapers; D, scum collector bar; E, sludge discharge hopper; F, inlet pipe; G, sludge discharge pipe; H, effluent discharge weir; J, drive motor [Courtesy of Templewood Hawksley Activated Sludge Ltd]

6.3 BIOFILTRATION

The liquid effluent from primary sedimentation tanks, termed 'settled sewage', is treated in one of two biological reactors, a biofilter or an activated sludge process. The biofilter (also known as the percolating, trickling or biological

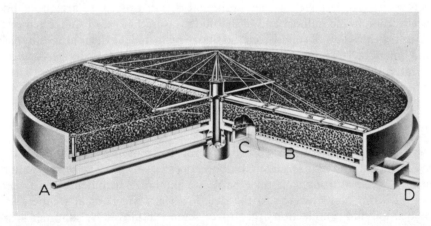

Figure 6.3 Sectional perspective view of a circular trickling filter showing rotating distributors and filter media. A, inlet pipe; B, underdrain blocks; C, effluent channel; D, outlet pipe [Courtesy of Dorr-Oliver Co. Ltd]

Figure 6.4 Irrigation of trickling filter with settled sewage

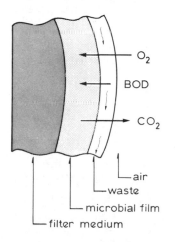

O_2

BOD

CO_2

air
waste
microbial film
filter medium

Figure 6.5 Schematic diagram
of BOD removal in a trickling
filter

filter and bacteria bed) is a circular or rectangular bed of coarse aggregate
(30–60 mm grading), usually 1·8 m deep (Figure 6.3). Settled sewage is distributed
over the bed and trickles down over the surface of the aggregate (Figure 6.4).
On these surfaces a *microbial film* develops and the bacteria, which constitute
most of this film, oxidize the sewage as it flows past (Figure 6.5). As the sewage
is oxidized the microbial film grows:

$$\text{settled sewage} + \text{oxygen} \xrightarrow{\text{bacteria}} \text{oxidized effluent} + \text{new bacterial cells}$$

Some of the new cells so formed are washed away from the film by the hydraulic
action of the sewage. These cells exert a high BOD and must be removed
before the effluent is finally discharged. This is achieved in secondary sedimentation tanks (often called 'humus' tanks; secondary tanks are basically similar

58

to primary tanks but without scum-skimming facilities). The clarified effluent is discharged usually to a river and the humus sludge pumped to the sludge treatment unit.

Recirculation

The proportion of voids in a filter which comprises 30–60 mm aggregate is 40–50 per cent of the gross volume. When a strong sewage is applied to such

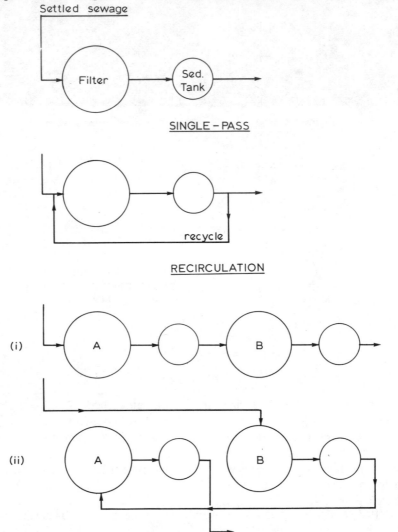

Figure 6.6 Flow diagrams for single-pass filters, recirculation and alternating double filtration (i) filter A leading and (ii) filter B leading

a filter excessive film growth occurs and this can lead to blockage of the filter and 'ponding' of the sewage on the surface of the filter. Experience has shown that if a final effluent of $BOD_5 < 20$ mg/l and $SS < 30$ mg/l is required the maximum loading that can be safely applied to a single-pass filter is 0·1 kg BOD_5/m^3d. However, if some of the clarified effluent is returned to the filter inlet (Figure 6.6), 2–3 times the normal loading for a single-pass filter can be applied with the production of a satisfactory effluent.

Recirculation serves not only to dilute the settled sewage but, more importantly, to reduce the rate of film growth and to increase the hydraulic stripping of the film; it also ensures that more of the available surface area is used for waste oxidation by providing a more uniform hydraulic distribution across the filter and also a more uniform vertical distribution of the microbial film. The recirculation ratio (ratio of the settled sewage flow to the recycle flow) is in the range 0·5–6·0, with 1·0 being the most common.

Alternating double filtration

An alternative way to achieve high loadings and satisfactory effluent quality is alternating double filtration (Figure 6.6). Two filters are operated in series. The first is loaded at a high rate (0·25 kg BOD_5/m^3d) and its effluent is, after settlement, applied to the second filter. The film growth in the first filter is very rapid and every 1–7 d (just before blockage would occur) the order of the filters is reversed and the film on the former first filter (now the second filter) disintegrates rapidly. Even though an extra humus tank and more pipework are required, the overall costs of alternating double filtration are less since the biofilters can handle considerably increased hydraulic and organic loadings.

Plastic media

Plastic media for biofilters have considerable advantages over traditional stone media: the voidage is > 90 per cent and their surface area per unit volume is 3–6 times higher. Although plastic media are most commonly used for the partial treatment of strong wastes in high rate biofiltration towers (Chapter 10), they can be used in conventional filters to produce effluents with $BOD_5 < 20$ mg/l and $SS < 30$ mg/l (Figures 6.7 and 6.8). Their major advantage is that the filter can be operated as a single-pass filter at high loadings (up to 0·5 kg

Figure 6.7 PVC filter medium, 'Flocor RC' [Courtesy of ICI Pollution Control Systems]

Figure 6.8 Trickling filter filled with random plastic medium [Courtesy of
ICI Pollution Control Systems]

BOD_5/m^3d) and yet still produce satisfactory effluents[6]; the cost of providing
the pumps and pipework for recirculation and alternating double filtration
are thus avoided.

6.4 ACTIVATED SLUDGE

Activated sludge is the conventional alternative to biofiltration. Settled sewage
is led to an aeration tank where oxygen is supplied either by mechanical agitation
(Figures 6.9 and 6.10) or by diffused aeration (Figures 6.11 and 6.12). The
bacteria which grow on the settled sewage are removed in a high-rate secondary
sedimentation tank. In order to maintain a high cell concentration (2000–8000
mg/l) in the aeration tank, most of the sludge is recycled from the sedimentation
tank to the aeration tank inlet. The sludge contains some inert solids but the
main components making up its loose, flocculent structure (Figure 6.13) are
living or 'active' bacteria and protozoa—hence the name 'activated sludge'.

Mechanism of BOD removal

In settled domestic sewage much of the BOD is associated with the small
suspended and colloidal solids; very little (< 5 per cent) of the BOD is due
to organic compounds present in true solution. There are two phases of BOD
removal by activated sludge. First there is a rapid initial removal by the entangle-

Figure 6.9 Surface aerator. Above, at rest; below, in action.
A, drive motor; B, cone; C, draft tube [Courtesy of Ames
Crosta Ltd]

62

Figure 6.10 Activated sludge tank with 14 surface aerators [Courtesy
of Ames Crosta Ltd]

ment of suspended solids within the gross sludge matrix and absorption of
colloidal material on to the floc surfaces. This phase is followed by a slow
progressive solubilization and oxidation of these waste compounds by the
bacteria present within the sludge flocs.

Activated sludge systems

The conventional system (Figure 6.14 (a)) is a plug flow reactor operated with
cell recycle. Oxygen is supplied at a uniform rate throughout the aeration tank,
even though the oxygen demand decreases along the length of the tank. To
overcome this waste, either the oxygen supply is progressively reduced along
the tank (this modification is termed *tapered aeration*) or the influent is added

Figure 6.11 Porous plates for diffused air supply
[Courtesy of Ames Crosta Ltd]

Figure 6.12 Activated sludge plant operated with diffused aeration [Courtesy
of Ames Crosta Ltd]

in several stages, a process known as *stepped aeration* (Figure 6.14 (b)). In all
of these systems the usual hydraulic retention time is 8–12 h at mean flow.
In the *contact-stabilization* process (Figure 6.14 (c)) the two phases of BOD
removal are separated. In the contact aeration tank a short time (0·5–1 h)
is provided for solids entanglement and adsorption; the solids are settled out

64

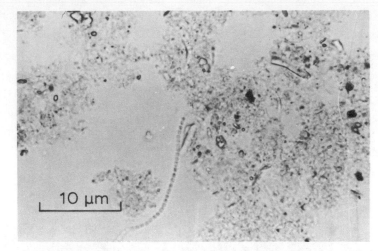

Figure 6.13 Micrograph of activated sludge

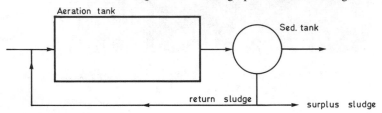

(a) CONVENTIONAL

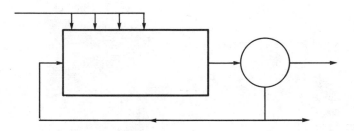

(b) STEPPED AERATION

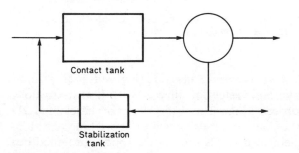

Figure 6.14 Flow diagrams for
activated sludge processes

(c) CONTACT STABILIZATION

with the activated sludge flows in the secondary sedimentation tank and then aerated for 2–4 h to permit their solubilization and oxidation and so re-activate the sludge. This modification is only suitable for wastes with most of their BOD associated with suspended and colloidal solids, but in such cases considerable reductions in both capital and running costs results as the total aeration volume is much smaller.

If the aeration period in the conventional plant process is extended to 18–48h the rate of sludge autolysis increases and substantially less sludge is produced. This principle of *extended aeration* is the basis of the oxidation ditch (Chapter 9).

6.5 SLUDGE TREATMENT

Primary and secondary sludges are most commonly treated together in a two-

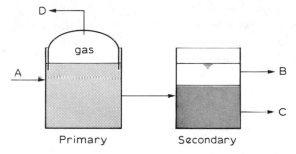

Figure 6.15 Conventional two-stage anaerobic digester. A, raw sludge inlet; B, supernatant liquor outlet; C, digested and thickened sludge outlet; D, methane draw-off pipe. In cold climates, the methane is used to heat the sludge in the primary tank; a pump is used to circulate sludge from the tank to a heat exchanger and back to the tank again

Figure 6.16 Anaerobic digesters [Courtesy of Ames Crosta Ltd]

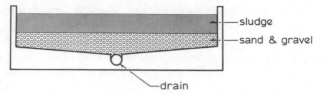

Figure 6.17 Sludge drying beds

stage anaerobic digester (Figures 6.15 and 6.16). The first stage is heated, if necessary, to 30–33 °C so that digestion can proceed more quickly; the methane gas released in the digestion process is commonly used to heat the digester contents. An alternative operating temperature is 50 °C which permits rapid digestion of the sludge by thermophilic bacteria. The second stage is a thickener for quiescent solids separation. The supernatant liquor has a BOD_5 of 5000–10 000 mg/l and is returned to the main works inlet for complete treatment. The digested sludge is in hot climates most advantageously placed on drying beds (Figure 6.17). When dry it may be sold as fertilizer.

6.6 PROBLEMS IN HOT CLIMATES

Operation and maintenance

Conventional sewage treatment relies heavily on electrical machinery—pumps, sludge scrapers, aerators—which requires considerable skill in installation, operation and maintenance. This skill, particularly in maintenance, is not

Figure 6.18 A neglected trickling filter at Molo, Kenya

readily available in many of the tropical developing countries; for example a survey conducted in Kenya[7] revealed that, outside Nairobi, none of the existing conventional works was working satisfactorily, the most common failure being associated with pumps and trickling filters (Figure 6.18).

Odour

In hot climate sewage can soon become malodorous ('septic') if insufficient oxygen is not made available to prevent the onset of anaerobic conditions. A higher level of odour can thus be expected in hot climates to come from primary sedimentation tanks which are by their nature designed for quiescent settling and not turbulent oxygenation. This odour is however almost always masked by the even higher odour release from the biofilters. Indeed so intense is the odour from low-rate biofilters that in many hot climates, for example California, activated sludge plants are used simply to overcome the biofilter odour problem.

Insect nuisance

The microbial film in biofilters is used as a breeding ground by various flies and midges. This is beneficial in that the larvae feed on the film and thus help to prevent ponding. However, although none of the species found in filters actually bites humans, their sheer numbers can be a severe nuisance in hot climates: clouds of *Psychoda* flies can effectively stop all human activity in and near a sewage treatment works.

6.7 DESIGN CRITERIA

Primary sedimentation

Overflow rate $= 30 \text{ m}^3/\text{m}^2$ d at peak flow

Retention time $= 2$ h at peak flow

Recent work has suggested that on the basis of overflow rate and retention time primary sedimentation tanks can be loaded at 2–5 times the rate given above and still achieve the same removal of SS.[8,9]

Biofiltration (stone media, 30–60 mm grading)

Organic loading $= 0.1 \text{ kg BOD}_5/\text{m}^3$ d

Hydraulic loading $= 0.5 \text{ m}^3/\text{m}^3$ d

Activated sludge (conventional)

Organic loading $= 0.5 \text{ kg BOD}_5/\text{m}^3$ d

Retention time $= 8\text{--}12$ h at mean flow (including return sludge flow)

Oxygen supply $= 1$ kg/kg BOD_5 applied

Secondary sedimentation

Overflow rate $= 40$ m^3/m^2 d at peak flow

Retention time $= 1\cdot5$ h at peak flow

Sludge treatment

Retention time $= 13$ d at 50 °C
$= 28$ d at 32 °C
$= 120$ d in unheated digesters in temperate climates

Drying bed area $= 0\cdot025\text{--}0\cdot1$ m^3/head

6.8 REFERENCES

1. 'Zambian sewage project completed', *Water & Waste Treatment*, December 1973, p.21.
2. Bradley, R. M. and Isaac, P. C. G., *Water Pollution Control*, **68**, 368 (1969).
3. Handley, J., *Water Pollution Control*, **73**, 230 (1974).
4. Clements, M. S. and Khattab, A. F. M., *Proceedings of the Institution of Civil Engineers*, **40**, 471 (1968).
5. Clements, M. S., *Surveyor*, 21 November 1969, p. 28.
6. Rogers, I., *Process Engineering*, August 1974, p.68.
7. Holland, R. J., personal communication, 1973.
8. Bradley, R. M., *Public Health Engineer*, (13), 4 (1974).
9. Tebbutt, T. H. Y., *Water Pollution Control*, **68**, 467 (1969); also *Water Research*, **9**, 347 (1975).

7

Waste Stabilization Ponds

7.1 INTRODUCTION

Waste stabilization ponds (Figure 7.1) are large shallow basins enclosed by earthen embankments in which raw sewage is treated by entirely natural processes involving both algae and bacteria. Since these processes are unaided by man (who merely allocates a place for their occurrence) the rate of oxidation is rather slow and as a result long hydraulic retention times are employed, 30–50 d not being uncommon. Ponds have considerable advantages (particularly as regarding costs and maintenance requirements and the removal of faecal bacteria) over all other methods of treating the sewage from communities of more than about 100 people. They are without doubt the most important method of sewage treatment in hot climates where sufficient land is normally available and where the temperature is most favourable for their operation.

Figure 7.1 A 0·8 ha facultative pond at Thika, Kenya

Their use is not of course restricted to hot climates: they are used at all latitudes, even as far north as Alaska. They are an important method of treatment in many industrialized countries—for example, nearly one-third of all municipal wastewater treatment plants in USA are stabilization ponds. There are three major types of pond: facultative, maturation and anaerobic ponds. A fourth type is the high-rate pond which is still largely experimental.

7.2 FACULTATIVE PONDS

These are the most common. They normally receive raw sewage or that which has received only preliminary treatment; they are, however, becoming increasingly used to treat the settled effluent from septic tanks and anaerobic pretreatment ponds. The term 'facultative' refers to a mixture of aerobic and anaerobic conditions and in a facultative pond aerobic conditions are maintained in the upper layers while anaerobic conditions exist towards the bottom.

Although some of the oxygen required to keep the upper layers aerobic comes from re-aeration through the surface, most of it is supplied by the photosynthetic activity of the algae which grow naturally in the pond where considerable quantities of both nutrients and incident light energy are available. Indeed, so profuse is the growth of algae that the pond contents are bright green in colour. The pond bacteria utilize this 'algal' oxygen to oxidize the organic waste matter. One of the major end-products of bacterial metabolism is carbon dioxide which is readily utilized by the algae during photosynthesis since their demand for it exceeds its supply from the atmosphere. Thus there is an association of mutual benefit ('symbiosis') between the algae and bacteria in the pond (Figure 7.2). Since photosynthesis is a light-dependent activity there is a diurnal variation in the amount of dissolved oxygen present in the pond (Figure 7.3) and a similar fluctuation in the level of the 'oxypause' (the point below the surface at which the dissolved oxygen concentration becomes zero) occurs. The pH of the pond contents also follows a daily cycle increasing with photosynthesis to a maximum which may be as high as 10. This happens because at peak demand algae remove CO_2 from solution more rapidly than

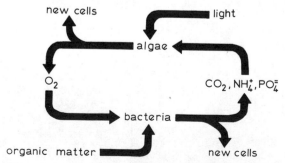

Figure 7.2 Symbiosis of algae and bacteria in stabilization ponds

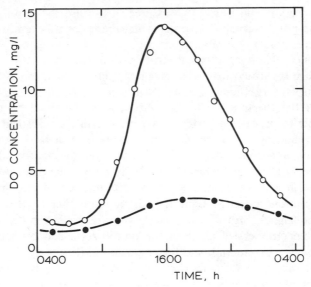

Figure 7.3 Diurnal variation in dissolved oxygen;
○, top 200 mm of pond; ●, 800 mm below surface

it is replaced by bacterial respiration: as a result the bicarbonate ions present dissociate to provide not only more CO_2 but also the alkaline hydroxyl ion which increases the pH value:

$$HCO_3^- \rightarrow CO_2 + OH^-$$

Mixing

Wind and heat are the two factors of major importance which influence the degree of mixing that occurs within a pond. Mixing fulfils a number of vital functions in a pond: it minimizes hydraulic short-circuiting and the formation of stagnant regions and it ensures a reasonably uniform vertical distribution of BOD, algae and oxygen. Mixing is the only means by which the large numbers of non-motile algae can be carried up into the zone of effective light penetration (the 'photic' zone); since the photic zone comprises only the top 150–300 mm of the pond, much of the pond contents would remain in permanent darkness if mixing did not occur. Mixing is also responsible for the transportation of the oxygen produced in the photic zone to the bottom layers of the pond. Good mixing thus increases the safe BOD load that can be applied to a pond.

The depth to which wind-induced mixing is felt is largely determined by the distance the wind is in contact with the water (the 'fetch'); an unobstructed contact length of about 100 m is required for maximum mixing by wind action. The importance of wind action is clearly demonstrated by the following admirably simple experiment which was conducted on a facultative pond in Zambia:[2]

a 2 m high fence with no openings was erected around the pond and within a few days the pond turned anaerobic; when the fence was removed aerobic conditions were rapidly re-established.

In the absence of mixing thermal stratification quickly occurs. The warm upper layers are separated from the cold lower layers by a thin static region of abrupt temperature change known as the *thermocline*. Non-motile algae settle through the thermocline to the darkness of the pond bottom where they are unable to produce any oxygen by photosynthesis; instead they exert an oxygen demand, with the result that conditions below the thermocline rapidly become anaerobic. Above the thermocline the motile algae move away from the hot surface waters (which may have a temperature > 35 °C) and usually form a dense layer about 300–500 mm below the surface (this layer of algae is an effective light barrier and the thermocline is usually just below the algae). Stratification is thus characterized by a substantial reduction in the numbers of algae in the photic zone and by a consequent reduction in oxygen production and hence waste stabilization.

The diurnal mixing pattern in a 1·5 m deep facultative pond in Lusaka, Zambia, has been thoroughly investigated and is probably typical of tropical and subtropical ponds:

1. In the morning, if there is any wind, there is a period of complete mixing in which the temperature is uniform throughout the pond but, owing to the absorption of radiation, gradually increases.
2. At some time, usually during a short lull in the wind, stratification develops abruptly and a thermocline forms. The temperature above the thermocline increases to a maximum and then decreases, while below the thermocline the temperature rapidly falls to a value approximately that of the earth and thereafter remains practically constant. A certain amount of mixing may take place above the thermocline.
3. In the afternoon and evening, a second period of mixing may be initiated as follows:
   (a) Above the thermocline, under quiescent wind conditions, the top layers lose their heat more rapidly than the bottom layers. The cooler top layers sink, inducing mixing, with the result that the temperature down to the thermocline remains approximately uniform but gradually decreases. The thermocline gradually sinks and, should the temperature above and below it become equal, with further cooling, mixing is initiated and sustained throughout the pond.
   (b) Under windy conditions, usually during the period of decreasing temperatures, the energy imparted by the wind to the water above the thermocline at some stage overcomes the stratification forces and progressively mixes the warmer and colder layer adjacent to the thermocline, causing it to be displaced downwards until the temperature is uniform throughout and the whole pond is in a state of mixing.[2]

Sludge layer

As the sewage enters the pond most of the solids settle to the bottom to form a sludge layer. At temperatures > 15 °C intense anaerobic digestion of the sludge solids occurs; as a result the thickness of the sludge layer is rarely more than about 250 mm and often much less. Desludging is only rarely required, once every 10–15 year. At temperatures > 22 °C the evolution of methane gas is sufficiently rapid to buoy sludge particles up to the surface where drifting sludge mats are formed. These must be removed (together with any other floating debris or scum) so that they do not prevent the penetration of light into the photic zone.

The soluble products of fermentation diffuse into the bulk liquid of the pond where they are oxidized further. The seasonal variation of the rate of fermentation (which increases approximately sevenfold with each 5 degC rise in temperature) explains why the BOD_5 in the pond often remains sensibly constant throughout the year in spite of the changes in temperature:

During summer the degradation rate is high and, from the theory, a low equilibrium BOD in the pond is established. However, the BOD load received from the sludge is high. During winter the degradation rate is low, establishing a relatively high equilibrium BOD in the pond, but a low BOD load is received from the sludge. The two processes, operating simultaneously, tend to cancel out, and decrease the cyclic variation of pond BOD.[2]

Depth

Depths < 1 m do not prevent the emergence of vegetation. This must be avoided as otherwise the pond becomes an ideal brooding ground for mosquitoes and midges. With depths > 1·5 m the oxypause is too near the surface with the result that the pond is predominantly anaerobic rather than predominantly aerobic. This is undesirable as the pond would have an unacceptably low factor of safety in normal operation and so be less able to cope with a fluctuating load or a sudden slug of heavy pollution. In arid climates eveporation rates are high and water losses should be minimized by increasing the depth to about 2 m and thus reducing the surface area. In cold climates (e.g. at high altitude) similar depths are used so as to preserve as much of the thermal energy of the influent sewage as possible. These considerations are usually more important in these extremes of climate than those concerned with the position of the oxypause.

Climatic influences

A hot climate is ideal for pond operation. Solar radiation is intense and, as a result, pond temperatures are high and there is more than an adequate intensity of light. The long daylight hours enable algal photosynthesis to occur for extended periods and so provide a reserve of dissolved oxygen for use during the night. There is however usually a month or more of seasonal cloud cover and, although light intensities during this time are sufficient for algal activity, the temperature falls to its annual minimum and it is this that limits both algal and bacterial growth. In order to ensure that the pond works satisfactorily at all times, it must be designed for the worst (i.e. coldest) conditions. The mean temperature of the coldest month is commonly used as the design temperature.

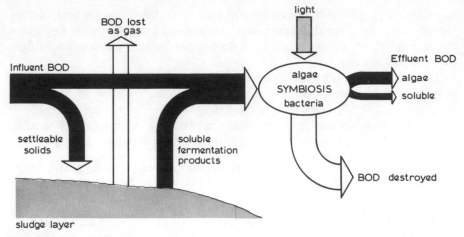

Figure 7.4 Pathways of BOD removal in facultative stabilization ponds [After Marais[1]]

Figure 7.4 shows in summary form, the principal pathways of BOD removal in facultative ponds.

7.3 MATURATION PONDS

Maturation ponds are used as a second stage to facultative ponds. Their main function is the destruction of pathogens. Faecal bacteria and viruses die off reasonably quickly owing to what is to them an inhospitable environment. The cysts and ova of intestinal parasites have a relative density of about 1·1 and as a result of the long retention times they settle to the bottom of the pond where they eventually die. The removal of BOD_5 in maturation ponds is small: two ponds in series, each with a retention time of 7 d, are required to reduce the BOD_5 from about 50–70 mg/l to < 25 mg/l (Section 7.9).

Maturation ponds are wholly aerobic and are able to maintain aerobic conditions at depths of up to 3 m. Most usually, however, the depth of a maturation lagoon is taken as the same as that of the associated facultative lagoon (1–1·5 m). This is advisable, as well as usually being convenient, since the destruction of viruses is better in shallow ponds than in deep ones.[3] The effectiveness of maturation ponds in removing pathogens is conveniently assessed by the removal of faecal coliforms. With proper design removals > 99·99 per cent can be achieved. In these circumstances no difficulty should be experienced in satisfying an effluent standard of < 5000 FC/100 ml (Section 3.4).

7.4 ANAEROBIC PRETREATMENT PONDS

These ponds are designed to receive such a high organic loading that they are completely devoid of dissolved oxygen. They are most advantageously used to pretreat strong wastes which have a high solids content. The solids settle to the bottom where they are digested anaerobically; the partially clarified supernatant liquor is discharged into a facultative pond for further treatment. The successful operation of anaerobic ponds depends on the delicate balance between the acid-forming bacteria and the methanogenic bacteria: thus a temperature > 15 °C is necessary and the pond pH must be > 6 (Section 3.3). Under these circumstances sludge accumulation is minimal: desludging which is required when the pond is half full of sludge, is necessary only every 3–5 years. At temperatures < 15 °C anaerobic ponds act merely as sludge storage basins.

This type of pond has in the past been unpopular with design engineers because of the fear of odour release and the extra maintenance required. The relationship between odour development and organic loading is now reasonably well understood so that this problem can usually be overcome at the design stage. The tremendous economies of land that are achieved by the use of anaerobic ponds (Section 7.14, problem 2) will often dictate their inclusion in large schemes (sewage flows $> 10\ 000$ m^3/d) where adequate maintenance facilities should be provided anyway.

7.5 HIGH-RATE PONDS

High rate ponds are designed to maximize algal growth[4] and so achieve high protein yields (Section 13.2). They are characterized by large area : volume ratios; depths are therefore shallow, 0·2–0·6 m. The pond contents need to be mixed once or twice each day to resuspend any settled solids, and removal of the algae from the final effluent is essential. An experimental pond in Thailand (depth $= 0·45$ m and retention time $= 1$ d), loaded with settled domestic sewage at a rate of 450 kg BOD$_5$/ha d, yielded 450 kg of algae/ha d; the final effluent had a BOD$_5 < 30$ mg/l.[5] The high-rate pond is undoubtedly a very efficient process but it has the disadvantage of requiring skilled personnel to operate and maintain the algae removal plant. Considerably more research and development work is required, however, before high-rate ponds can be recommended for general use.

7.6 ADVANTAGES OF PONDS

The major disadvantage of ponds is that they require much larger areas of land than other forms of sewage treatment. However, in many countries, especially tropical developing countries, this is rarely a disadvantage of any importance since sufficient land is normally available at relatively low cost. Some advantages of ponds are:

(1) *They can achieve any required degree of purification at the lowest cost and with the minimum of maintenance by unskilled operators.*

Figure 7.5 shows annual cost ranges (maintenance costs + capital costs discounted at 6 per cent p.a. for 20 years) for various methods of sewage treatment in India:[6] waste stabilization ponds are clearly the cheapest form of treatment.

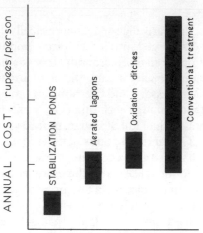

Figure 7.5 Annual costs of sewage treatment in India [From data given by CPHERI[6]]

Maintenance requirements are minimal: regular cutting of the grass embankments and the removal of any floating scum from the pond surface are all that is required. In spite of their simplicity these tasks are extremely important and operator training and supervision is essential. Labour requirements have been estimated as follows:[7]

Population served	Supervisors	Labourers
5 000	—	2
10 000	—	3
50 000	1	6
100 000	2	8

(2) *The removal of pathogens is considerably greater than that in other methods of sewage treatment.*

The effluent from a series of three ponds usually contains < 5000 FC/100 ml whereas the final effluent from a conventional works (humus tank effluent) typically contains about 5 000 000 FC/100 ml. Cysts and ova of intestinal parasites, which are commonly present in conventional effluents, are not found in maturation pond effluent. The pond habitat is fortunately unsuitable for the growth of the snail hosts of parasitic trematode worms such as *Schistosoma* spp. and *Clonorcha sinensis*.[8]

(3) *They are well able to withstand both organic and hydraulic shock loads.*

The long retention times (20–30 d in facultative ponds receiving raw sewage) ensure that there is always sufficient dilution available for short shock overloads.

(4) *They can effectively treat a wide variety of industrial and agricultural wastes.*

Wastes which are readily biodegradable (such as those from dairies, slaughter houses and food processing factories) have been successfully treated with domestic sewage in facultative ponds.[9] Anaerobic ponds are particularly advantageous for strong wastes that are to be treated alone. Some agricultural wastes are very strong indeed, for example:

Piggery wastes	20 000 mg BOD_5/l
Meat packing wastes	1000–3000 mg BOD_5/l
Dairy wastes	800–15 000 mg BOD_5/l

Pretreatment in anaerobic ponds prior to discharge in a public sewer may be required by the local authority in order to prevent overloading of existing sewage treatment facilities.

The high pH of the pond causes toxic heavy metals to precipitate as the hydroxide and so be removed into the sludge layer. Experiments conducted in Israel[10] showed that a heavy metal concentration of 30 mg/l (6 mg/l each of cadmium, hexavalent chrome, copper, nickel and zinc) did not affect pond performance, but that 60 mg/l did. Thus a facultative pond normally operating above pH 8 should be able to tolerate the heavy metals in municipal sewage for a considerable period of time before their accumulation in the sludge layer would affect its performance.

(5) *They can easily be designed so that the degree of treatment is readily altered.*

By designing the pond outlet structure (Section 7.12) so that the top water level can be varied, the retention time and hence the degree of treatment can be altered. This can be a useful device when a pond receives a seasonal waste (e.g. a food-processing waste) in addition to its normal sewage flow.

(6) *The method of construction is such that, should at some future date the land be required for some other purpose, it is easily reclaimed.*

All that is required is the removal of the pond inlet and outlet structures and of the paving slabs at top water level. The ground should then be levelled. The sale of former pond land will usually yield a substantial real estate profit to the municipality.

(7) *The algae produced in the pond are a potential source of high-protein food which can be conveniently exploited by fish farming.*

Fish have been successfully grown in maturation ponds (Section 13.2). The sale of fish can bring in a substantial revenue and so reduce the running costs of the treatment works. Ducks may also be reared on maturation ponds.

Pond sizes

The advantages of ponds are such that they are not infrequently used for very large sewage flows. Large flows of course require large ponds. Some examples are:

Auckland, New Zealand	530 ha	210 000 m^3/d
Melbourne, Australia	310 ha	350 000 m^3/d
Stockton, California	250 ha	250 000 m^3/d

It is noteworthy that these examples are all from industralized countries. In hot climates ponds should always be considered the first method of choice for sewage treatment. Indeed, a very good case must be made for *not* using them.

7.7 TYPICAL POND LAYOUTS

It has been frequently observed that the effluent from a series of ponds is of better quality than that from a single pond of the same size. This is because although the hydraulic regime in individual ponds is closer to complete mixing than it is to plug flow, the overall performance of a series of ponds approximates that of a plug flow reactor. To illustrate this point, consider a series of n identical ponds and assume that each is a completely mixed reactor in which BOD removal follows first order kinetics. A mass balance of BOD around the mth pond (Figure 7.6) yields an equation similar to equation 4.7:

$$\frac{L_m}{L_{m-1}} = \frac{1}{1 + k_1 t^*} \tag{7.1}$$

For the whole series of n ponds:

$$\frac{L_e}{L_i} = \frac{1}{(1 + k_1 t^*)^n} \tag{7.2}$$

If ϕ is the total retention time in the series of ponds, then:

$$\phi = n t^* \tag{7.3}$$

and

$$\frac{L_e}{L_i} = \frac{1}{(1 + k_1 t^*)^{\phi/t^*}} \tag{7.4}$$

In the limit as $t^* \to 0$ (and, if ϕ is to remain the same, as $n \to \infty$):

$$\frac{L_e}{L_i} = \lim_{t^* \to 0} \left(\frac{1}{(1 + k_1 t^*)^{\phi/t^*}} \right)$$

i.e.

$$\frac{L_e}{L_i} = e^{-k_1 \phi} \tag{7.5}$$

which is the equation for a plug flow reactor of retention time ϕ (compare equation 4.9). This shows that a plug flow reactor can be considered as a series of an infinitely large number of infinitely small completely mixed reactors. Since plug flow is the most efficient hydraulic regime (Section 4.2), equation 7.5 demonstrates that a series of small ponds is more efficient than a single large pond.

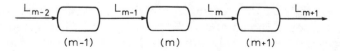

Figure 7.6

Marais' theorem[11]

This theorem states that *maximum efficiency in a series of ponds is achieved when the retention time in each pond is the same.* The theorem is proved here for a series of two ponds, but the proof is applicable in principle for any number of ponds.

Let the retention time in the first pond be t_1^* and in the second pond t_2^* and let $t_1^* + t_2^* = \phi$. The effluent BOD_5 L_e from the second pond is given by the equation:

$$L_e = \frac{L_i}{(1 + k_1 t_1^*)(1 + k_1 t_2^*)}$$

L_e will be a minimum when $(1 + k_1 t_1^*)(1 + k_1 t_2^*)$ is a maximum.

$$\text{Let} \quad y = (1 + k_1 t_1^*)(1 + k_1 t_2^*)$$

$$\therefore \quad y = (1 + k_1 t_1^*)[1 + k_1(\phi - t_1^*)]$$

$$= 1 + k_1 \phi + k_1^2 \phi t_1^* - (k_1 t_1^*)^2$$

$$\therefore \quad \frac{dy}{dt_1^*} = k_1^2 \phi - 2 k_1^2 t_1^*$$

$$= 0, \text{ for a maximum}$$

$$\therefore \quad t_1^* = \phi/2$$

$$\therefore \quad t_1^* = t_2^* \text{ for maximum } y \text{ and hence minimum } L_e.$$

That y is a maximum when $t_1^* = t_2^*$ and not a minimum is verified by considering the second differential coefficient:

$$d^2 y/d(t_1^*)^2 = -2 k_1^2$$

Since this is negative, y is a maximum.

The two alternative pond layouts that are acceptable in tropical countries are shown in Figure 7.7. It is extremely bad engineering practice to build only a single facultative pond: maturation ponds are of vital importance as they are responsible for pathogen removal. A multi-pond system comprising an anaerobic, a facultative and three or more maturation ponds, each with a retention time of 5 d is recommended[12] as the minimum treatment required when the final effluent is to be used for unrestricted irrigation (Section 7.14, example 3).

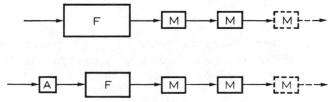

Figure 7.7 Pond layouts. A, anaerobic pond; F, facultative pond; M, maturation pond

7.8 DESIGN PRELIMINARIES

The design of waste stabilization ponds is part rational and part empirical. The depth is selected with due regard to the site and to the considerations given above. The range of depths most commonly used for each type of pond is as follows:

facultative ponds	1–1·5 m
maturation ponds	1–1·5 m
anaerobic ponds	2–4 m

The length and breadth are usually in the ratio 2–3 to 1; they are obtained from the mid-depth area as shown in Figure 7.8. The mid-depth area (which

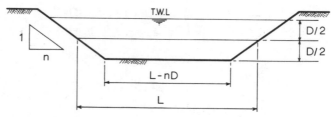

Figure 7.8 Calculation of dimensions of pond bottom from those derived from mid-depth area

is what the surface and base areas would be if the pond had vertical sides) is given by:

$$A = Qt^*/D \qquad (7.6)$$

where Q = volumetric flow rate, m^3/d

t^* = retention time, d

D = pond depth, m.

The areal or surface BOD_5 loading λ_s is the weight of BOD_5 applied per unit area per day. The weight (g) of BOD_5 applied to the pond (A, m^2) each day is L_iQ. Therefore, expressing λ_s in its usual units of kg/ha d:

$$\lambda_s = \frac{10^{-3}L_iQ \ (kg/d)}{10^{-4}A \ (ha)}$$

$$\lambda_s = 10L_iQ/A \qquad (7.7)$$

Volumetric, rather than surface, loading rates are sometimes used, particularly with anaerobic ponds. The volumetric loading λ_v is the weight of BOD_5 applied per unit volume per day. Therefore, expressing λ_v in g/m^3d:

$$\lambda_v = L_iQ/AD \qquad (7.8)$$

or, since AD/Q = retention time, t^*:

$$\lambda_v = L_i/t^* \tag{7.9}$$

In facultative ponds, λ_v is usually in the range 15–30 g/m^{3}d, whereas in anaerobic ponds, where it is a good criterion for odour development, values are > 100 g/m^{3}d.

7.9 DESIGN OF FACULTATIVE PONDS

Note. Design procedures for facultative ponds, other than those described below, are discussed in Appendix 4.

First order kinetics

The simplest approach to the rational design of facultative ponds is to assume that they are completely mixed reactors in which BOD$_5$ removal follows first order kinetics.[13] Equation 4.7 is therefore applicable:

$$\frac{L_e}{L_i} = \frac{1}{1 + k_1 t^*} \tag{4.7}$$

Rearranging:

$$t^* = \left(\frac{L_i}{L_e} - 1\right)\frac{1}{k_1} \tag{7.10}$$

Substituting for t^* in equation 6.6 gives:

$$A = \frac{Q}{Dk_1}\left(\frac{L_i}{L_e} - 1\right) \tag{7.11}$$

Work in South Africa[13,14] has suggested that in order to maintain the pond contents predominantly aerobic (rather than predominantly anaerobic) L_e should be in the range 50–70 mg/l for pond depths of 1–1·5 m. The value of k_1 is about 0·3 d^{-1} at 20 °C and its variation with temperature is described by an equation similar to equation 4.8:[15]

$$k_{1(T)} = 0\cdot3(1\cdot05)^{T-20} \tag{7.12}$$

Substitution of this equation into equation 6.10 and selection of L_e as 60 mg/l gives the following design equation for A:

$$A = \frac{Q(L_i - 60)}{18\,D(1\cdot05)^{T-20}} \tag{7.13}$$

The temperature should be taken as the mean temperature of the coldest month.

Dispersed flow

The Wehner-Wilhelm equation (equation 4.11) for first order BOD$_5$ removal in dispersed flow reactors has been used[16] for the design of facultative ponds but, although in principle it forms a more rational basis for design, it is not in practice as useful as the equations based on complete mixing. The reasons for this are twofold. Firstly, for the same efficiency a completely mixed reactor requires more volume than a dispersed flow reactor, so that the assumption of complete mixing provides a factor of safety in the design since dispersion numbers are always $< \infty$. Secondly, reported values of the rate constant k_1 have been usually calculated by inserting measured values of L_i, L_e, Q and T into either equation 4.7 or 4.9; indeed this is the only means of calculation possible whenever (as is normally the case) the dispersion number is not measured as well. This point is illustrated by the following example.

Example. For a pond having $L_i = 600$ mg/l, $L_e = 60$ mg/l and $t^* = 30$ d, calculate k_1 assuming (a) complete mixing, (b) dispersed flow conditions with $\delta = 1$, and (c) plug flow.
    Solution: (a) From equation 4.7, $k_1 = [(L_i/L_e) - 1]/t^* = [(600/60) - 1]/30 = 0.30$ d^{-1}.
(b) From Figure 4.3, $k_1 t^* = 5$ for $\delta = 1$ and a removal of 90 per cent. Hence $k_1 = 5/30 = 0.17$ d^{-1}
(c) From equation 4.9, $k_1 = - [\ln(L_e/L_i)]/t^* = - [\ln(60/600)]/30 = 0.08$ d^{-1}.

McGarry and Pescod empirical procedure[17]

An analysis of operational data from many facultative ponds all over the world (Figure 7.9) showed that the maximum BOD$_5$ surface loading that could be applied to a facultative pond before it failed (i.e. became completely anaerobic) was related to the mean monthly ambient air temperature (this being taken as the most convenient measure of climate for which records exist) as follows:

$$\hat{\lambda}_s = 11.2 \, (1.054)^T \qquad (7.14)$$

where $\hat{\lambda}_s =$ maximum BOD$_5$ loading, kg/ha d
    $T =$ temperature, °F.
However, ponds are not normally designed to operate just at their point of failure. For design purposes therefore the introduction of a safety factor would be required; for example equation 7.14 could be modified to:[18]

$$\lambda_s = 7.5 \, (1.054)^T \qquad (7.15)$$

where $\lambda_s =$ design loading, kg/ha d.
An alternative design equation for λ_s is the straight line relationship shown in Figure 7.9:

$$\lambda_s = 20 \, T - 120 \qquad (7.16)$$

where T is in °C.
The design equation for A is then simply obtained from equation 7.7 as:

$$A = \frac{L_i Q}{2T - 12} \qquad (7.17)$$

    The degree of BOD$_5$ removed in facultative ponds was found to be related to the BOD$_5$ loading (Figure 7.10):[17]

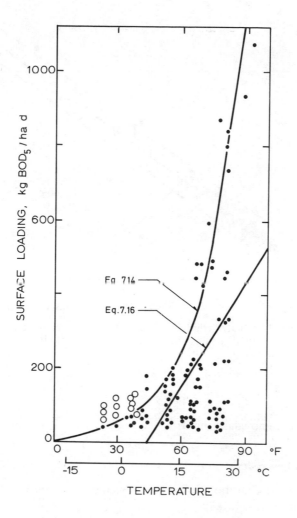

Figure 7.9 Variation of maximum permissible and design loadings on facultative ponds with mean air temperature. The shape of the curve of equation 7.14 is largely influenced by (1) the occurrence of eleven data points of pond failure clustered about 0 °C and (2) the relatively few data points of successful operation at atypically high loadings (> 600 kg/ha d). Equation 7.16, if restricted to temperatures between 15 °C and 30 °C, is probably as realistic a design equation as any other and it has the advantage of simplicity of form [Adapted from McGarry and Pescod[17]]

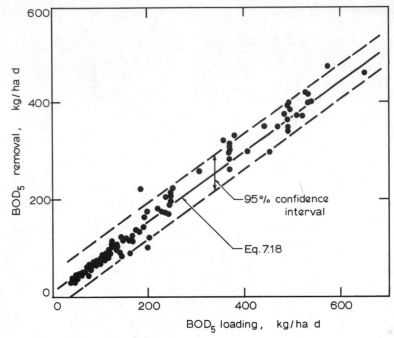

Figure 7.10 BOD removal in facultative ponds as a function of BOD loading [From McGarry and Pescod[17]]

$$\lambda_r = 0{\cdot}725\lambda_s + 10{\cdot}75 \tag{7.18}$$

where λ_r = BOD removal, kg/ha d.

Indian empirical procedure[19]

Experience of pond operation in India has yielded a design which relates the permissible loading to latitude:

$$\lambda_s = 375 - 6{\cdot}25L \tag{7.19}$$

where L = latitude (range in India: 8–36 °N).

Inasmuch as latitude is some measure of temperature this design procedure is basically similar to the McGarry and Pescod procedure.

7.10 DESIGN OF MATURATION PONDS

In order to produce an effluent with a $BOD_5 < 25$ mg/l it has been found that two maturation ponds in series, each with a retention time of 7 d, are required.[13] This assumes that the BOD_5 of the influent (i.e. of the facultative pond effluent) is less than about 75 mg/l.

Bacterial reduction

The reduction of faecal bacteria in a pond (anaerobic, facultative or maturation) has been found to follow first order kinetics. The appropriate version of equation 4.7 is:

$$N_e = \frac{N_i}{1 + K_b t^*} \tag{7.20}$$

where N_e = number of FC/100 ml of effluent
$\quad N_i$ = number of FC/100 ml of influent
$\quad K_b$ = first order rate constant for FC removal, d^{-1}.
For n ponds in series equation 7.20 becomes:

$$N_e = \frac{N_i}{(1 + K_b t_1^*)(1 + K_b t_2^*)\ldots(1 + K_b t_n^*)} \tag{7.21}$$

where t_n^* = retention time in the nth pond.
The value of K_b is extremely temperature sensitive; it is given by the equation:[11]

$$K_{b(T)} = 2\cdot6\,(1\cdot19)^{T-20} \tag{7.22}$$

where $K_{b(T)}$ = the value of K_b at T °C.
A reasonable design value of N_i is 4×10^7 FC/100 ml; this is slightly higher than average values normally found in practice.

The best design procedure is to calculate the retention time in the facultative pond and determine the value of N_e from equation 7.20 which results from having two maturation ponds each with $t^* = 7$ d. If this value of N_e is unacceptable, then three or more maturation ponds each with $t^* = 5$ d are chosen and N_e recalculated on this basis (Section 7.14, example 3).

Although faecal coliforms are commonly used to indicate the removal of faecal organisms in a series of ponds, there is evidence that some pathogenic bacteria do not die off as quickly as do faecal coliforms—for example, a salmonella was found[2] to have a K_b value of $0\cdot8$ d^{-1} in the same pond as faecal coliforms with a K_b value of $2\cdot0$ d^{-1}. Also drug-resistant coliforms are known to die off more slowly than those without resistance genes.[20]

7.11 DESIGN OF ANAEROBIC PONDS

Provided that the pH is > 6, BOD reduction in anaerobic ponds is a function of temperature (increasing with increasing temperatures > 15 °C) and of BOD loading (the higher the loading, the greater the reduction). Unfortunately, however, there are insufficient field data to obtain a meaningful correlation between BOD reduction and these variables that could be used confidently for design. An examination of the operational results of anaerobic ponds in Israel, Africa and Australia has suggested the following design values of BOD_5 reduction for varying retention times at temperatures > 20 °C:[15]

Retention time (d)	BOD_5 reduction (%)
1	50
2·5	60
5	70

These values are slightly less than those found in practice and therefore lead to a conservative design. For temperatures in the range 15–20 °C the BOD_5 reduction may be estimated as 10–20 per cent less than the figures given above.

The *optimum retention time* is 5 d. Ponds operated with retention time > 5 d have been found to be facultative rather than anaerobic.[21] Retention times < 5 d are of course possible but are not recommended because (1) the risk of odour release is greater; (2) the interval between successive desludging operations is shorter; (3) the bacteriological quality of the final effluent is poorer; and (4) the BOD removal is smaller.

Desludging frequency

The rate of sludge accumulation is approximately 0·03–0·04 m^3/hd year and desludging is required when the pond is half-full of sludge.[22] This occurs every *n* years where *n* is given by:

$\frac{1}{2}$(pond volume, m^3) × (sludge accumulation rate, m^3/hd yr) × (population)

For the purpose of design the rate of sludge accumulation may be estimated as 0·04 m^3/hd yr.

Odour release and control

The release of objectionable odours from anaerobic ponds occurs when the volumetric loading on the pond is > 400 g BOD_5/m^3d.[14] Thus even for a very strong sewage (say, BOD_5 = 1000 mg/l) odour release is unlikely to be a problem when a retention time of 5 d is employed. The presence of industrial or agricultural wastes, particularly those with high concentrations of sulphate, may cause odour release. Odour control then becomes necessary and this may be achieved by:

(1) Raising the pH of the pond to about 8 so that most of the sulphide—formed by the bacterial reduction of sulphate—will exist as the odourless bisulphide ion (HS^-); under these conditions the release of the malodorous hydrogen sulphide gas (H_2S) is virtually non-existent.
(2) Recirculating the effluent from the facultative or maturation ponds to the anaerobic pond inlet in the ratio 1 to 6 (1 volume of effluent to 6 volumes of raw sewage).[23]

7.12 FACILITIES DESIGN

Pond geometry

The hydraulic characterictics of rectangular ponds have been found[24] to be

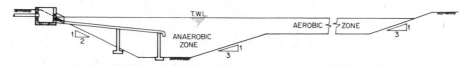

Figure 7.11 Deepened inlet zone in facultative pond

superior to those of square and circular ponds and those with irregular geometry. Length to breadth ratios of 2 to 1 and 3 to 1 are most frequently used. However, long thin ponds have lower dispersion numbers and their performance may therefore be expected to be better; if the site topography is suitable for a long thin pond it is advisable to deepen the pond near the inlet to create an anaerobic zone for solids deposition and digestion (Figure 7.11).

Parallel units

In order to facilitate maintenance it is advisable to divide the sewage flow into two or more parallel streams which are then treated in individual series of ponds. The actual number of parallel units depends on the magnitude of the flow and the topography of the site. The latter is important because in order to minimize construction costs it is necessary to minimize earthworks by balancing the volumes of cut and fill. As a guide the flow through each series of ponds should be < 5000 m^3/d and preferably < 2500 m^3/d.

Pond base

The bottom of the pond should be impermeable. Although the sludge layer seals up small pores in the soil, sealing of the base is necessary in coarse permeable soils when the seepage rate is more than about 10 per cent of the inflow. Polythene sheeting (sandwiched between two 100 mm layers of selected fill) and linings of clay, bitumen and asphalt have all been used successfully.

Embankment

An embankment slope of 1 in 3 is usually satisfactory in most soil conditions. If steeper slopes are used, their stability should be established by standard soil mechanics procedures.

Erosion of the embankment by surface wave action is avoided by placing precast concrete slabs at the top water level. The slabs stop vegetation growing down the banks and so prevent the breeding of mosquitoes; they also make maintenance easier (Figures 7.12 and 7.13).

Inlet structure

All pond systems (except possibly the very smallest) should have a venturi

Figure 7.13 ...comparing what happens in their absence

Figure 7.12 The advantages of setting paving slabs at top water level on pond embankments are readily appreciated by....

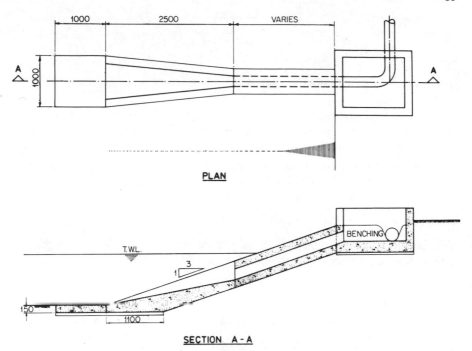

Figure 7.14 Inlet chute for facultative and maturation ponds

or Parshall flume to measure the inflow and a vee-notch to measure the final outflow. This enables the performance of the pond to be checked in a meaningful manner and will show when the hydraulic design capacity has been reached.

A pipe supported on piers is used to convey the sewage into anaerobic and facultative ponds. In order to reduce the amount of scum the pipe should discharge below the pond surface and in order to prevent the formation of a sludge bank the end of the pipe should be sited up to about 25 m (depending on the pond dimensions) away from the embankments. An alternative arrangement which is satisfactory for facultative ponds which receive comminuted or settled sewage is the inlet chute shown in Figure 7.14; this arrangement is also suitable for maturation ponds.

Interpond connection

The simplest (and cheapest) form of interpond connection is the piping arrangement shown in Figure 7.15. When the difference in the top water level of the two ponds is > 1 m the pipe should discharge into the inlet chute shown in Figure 7.14.

Interpond connections should be designed to pass a flow of $1.5Q$ where Q is the average daily flow. Flows > $1.5Q$ are then attenuated in the pond and in this way the occurrence of floods flashing through a series of ponds is prevent-

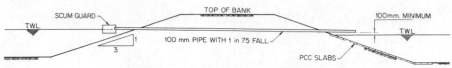

Figure 7.15 Simple interpond connection

ed. In order to minimize hydraulic short-circuiting the inlet and outlet to each pond should be located in diagonally opposite corners. If the length to breadth ratio is less than 2 to 1 multiple inlets and outlets should be provided. Wind-induced short-circuiting can be reduced by siting the inlet and outlet on the pond diagonal normal to the direction of the prevailing wind.

Outlet structure

The interpond connection shown in Figure 7.15 is suitable if it discharges into a chamber containing a vee-notch.

7.13 REFERENCES

1. Marais, G. v. R., in *Proceedings of the Second International Symposium for Waste Treatment Lagoons, Kansas City, Mo.*, 1970.
2. Marais, G. v. R., *Bulletin of the World Health Organization*, **34**, 737 (1966).
3. Malherbe, H. H. and Coatzee, O. J., *CSIR Research Report No. 242*, Council for Scientific and Industrial Research, Pretoria, 1965.
4. Oswald, W. J. and Gotaas, H. B., *Transactions of the American Society of Civil Engineers*, **81**, 686 (1955).
5. McGarry, M. G., in *Water Supply and Waste-Water Treatment in Developing Countries* Asian Institute of Technology, Bangkok, 1970.
6. *Cost of Sewage Treatment (Technical Digest No. 10)*, Central Public Health Engineering Research Laboratory, Nagpur, 1970.
7. Arceivala, S. J. *et al.*, in *Low Cost Waste Treatment*, Central Public Health Engineering Research Institute, Nagpur, 1972.
8. Hodgson, H. T., *Journal of the Water Pollution Control Federation*, **36**, 51 (1964).
9. Olson, O. O., *Journal of the Water Pollution Control Federation*, **40**, 414 (1968).
10. Moshe, M. *et al.*, *Water Research*, **6**, 1165 (1972).
*11. Marais, G. v. R., *Journal of the Environmental Engineering Division, American Society of Civil Engineers*, **100**, 119 (1974).
12. Mara, D. D., *Journal of the Environmental Engineering Division, American Society of Civil Engineers*, **101**, 296 (1975).
13. Marais, G. v. R. and Shaw, V. A., *Transactions of the South African Institute of Civil Engineers*, **3**, 205 (1961).
*14. Meiring, P. G. *et al.*, *CSIR Special Report WAT 34*, Council for Scientific and Industrial Research, Pretoria, 1968.
*15. Mara, D. D., *Design Manual for Sewage Lagoons in the Tropics*, East African Literature Bureau, Nairobi, 1975.
16. Thirumurthi, D., *Journal of the Sanitary Engineering Division, American Society of Civil Engineers*, **95**, 11 (1969).
17. McGarry, M. G. and Pescod, M. B., in *Proceedings of the Second International Symposium on Waste Treatment Lagoons, Kansas City, Mo.*, 1970.
18. Mara, D. D., *Water Research*, **9**, 595 (1975).

19. *Oxidation Ponds* (*Technical Digest No. 12*), Central Public Health Engineering Research Institute, Nagour, 1970.
20. Grabow, W. O. K. *et al.*, *Water Research*, **7**, 1589 (1973).
21. Parker, C. D., *Journal of the Water Pollution Control Federation*, **34**, 149 (1962).
22. Vincent, L. J. *et al.*, in *Proceedings of a Symposium on Hygiene and Sanitation in relation to Housing*, Commission for Technical Cooperation in Africa, London, 1963.
23. Van Eck, H. and Simpson, D. F., *Journal and Proceedings of the Institute of Sewage Purification*, (3), 251 (1966).
24. Shindala, A. and Murphy, W. C., *Water & Sewage Works*, (10), 394 (1969).
*Recommended reading

Further reading

Arceivala, S. J. *et al.*, *Waste Stabilization Ponds: Design, Contraction and Operation in India*, Central Public Health Engineering Research Unit, Nagpur, 1970.
Gloyna, E. F., *Waste Stabilization Ponds*, World Health Organization, Geneva, 1971.
Sless, J. B., 'Biological and chemical aspects of stabilization pond design', *Reviews on Environmental Health*, **1** (4), 327–354 (1974).
Watson, J. L. A., Oxidation ponds and use of effluent in Israel, *Proceedings of the Institution of Civil Engineers*, **22**, 21–40 (1962).

7.14 DESIGN EXAMPLES

1. *Design a facultative and maturation pond system to treat 10 000 m^3/d of domestic sewage which has a BOD_5 of 630 mg/l. The design temperature is 20 °C and the required effluent standards are: $BOD_5 < 25$ mg/l and FC < 5000/100 ml.*
Solution

Facultative pond: Choose $L_e = 60$ mg/l and $D = 1.2$ m and calculate A from equation 7.13:

$$A = \frac{Q(L_i - 60)}{18D(1.05)^{T-20}}$$

$$= \frac{10\,000(630-60)}{18 \times 1.2 \times 1} = 264\,000 \text{ m}^2$$

Calculate organic loading from equation 7.7:

$$\lambda_s = \frac{10QL_i}{A}$$

$$= \frac{10 \times 10\,000 \times 630}{264\,000} = 240 \text{ kg/ha d}$$

Check permissible loading from equation 7.16:

$$\lambda_s = 20\,T - 120$$

$$= (20 \times 20) - 120 = 280 \text{ kg/ha d}$$

Therefore the design can be accepted as satisfactory.

Maturation ponds: Check bacteriological quality of effluent from two ponds in series each with $t^* = 7$ d.

$$\text{Retention time in facultative pond} = \frac{\text{volume}}{\text{flow}} = \frac{264\,000 \times 1.2}{10\,000} = 32 \text{ d}$$

Therefore from equation 7.21 (assuming $K_b = 2.6$ at 20 °C and $N_i = 4 \times 10^7$):

$$N_e = \frac{N_i}{(1 + K_b t^*_{fac})(1 + K_b t^*_{mat})^2}$$

$$= \frac{4 \times 10^7}{[1 + (2.6 \times 32)][1 + (2.6 \times 7)]^2}$$

$$= 1300 \text{ FC/100 ml} \qquad (satisfactory)$$

Choose $D = 1.2$ m and calculate A from equation 7.6:

$$A = \frac{Qt^*}{D}$$

$$= \frac{10\ 000 \times 7}{1.2} = 58\ 000 \text{ m}^2$$

Remarks: After preliminary treatment the flow should be split into four streams. Thus the pond system would comprise four facultative ponds of 6·6 ha each and eight maturation ponds of 1·5 ha each.

2. *Repeat problem 1 incorporating anaerobic pretreatment.*
Solution
Anaerobic pond: Choose $D = 3$ m and $t^* = 5$ d; calculate A from equation 7.6:

$$A = \frac{Qt^*}{D}$$

$$= \frac{10\ 000 \times 5}{3} = 13\ 400 \text{ m}^2$$

Facultative pond: Influent $BOD_5 = 0.3 \times 630 = 190$ mg/l (70 per cent reduction in anaerobic pond). Choose $D = 1.5$ m and calculate A from equation 7.13:

$$A = \frac{Q(L_i - 60)}{18\ D(1.05)^{T-20}}$$

$$= \frac{10\ 000 \times 190}{18 \times 1.5} = 70\ 000 \text{ m}^2$$

Maturation ponds: Choose $D = 1.5$ m and $t^* = 7$ d; calculate A from equation 7.6:

$$A = \frac{Qt^*}{D}$$

$$= \frac{10\ 000 \times 7}{1.5} = 47\ 000 \text{ m}^2$$

Check bacteriological quality:

$$t^*_{fac} = \frac{70\ 000 \times 1.5}{10\ 000} = 10.5 \text{ d}$$

Therefore from equation 7.21:

$$N_e = \frac{N_i}{(1 + K_b t^*_{an})(1 + K_b t^*_{fac})(1 + K_b t^*_{mat})^2}$$

$$= \frac{4 \times 10^7}{[1 + (2.6 \times 5)][1 + (2.6 \times 10.5)][1 + (2.6 \times 7)]^2}$$

$$= 280 \text{ FC/100 ml} \qquad (satisfactory)$$

Remarks: Table 7.1 gives the areas and retention times required for the two solutions, one with and the other without anaerobic pretreatment.

Table 7.1

	With pretreatment		Without pretreatment	
	A	t*	A	t*
Anaerobic pond	1·3	5	—	—
Facultative pond	7·0	11	26·4	32
Maturation ponds	9·4	14	11·6	14
Totals	17·7 ha	29 d	37·0 ha	46 d

Anaerobic pretreatment thus results in savings of 52 per cent in pond area and 33 per cent in retention time. This example clearly quantifies the statement of Professor Marais[1] that **'anaerobic pretreatment is so advantageous that the first consideration in the design of a series of ponds should always include the possibility of anaerobic pretreatment'**.

3. *Design a stabilization pond system to treat 1000 m^3/d of domestic sewage which has a DOD_5 of 500 mg/l to the WHO standard for unrestricted irrigation (FC < 100/100 ml). The design temperature is 17 °C.*
Solution
From equation 7.22 the value of K_b at 17 °C is:

$$K_b = 2·6(1·19)^{-3} = 1·55 \text{ d}^{-1}$$

Consider a series of n ponds, each with $t^* = 5$ d; from equation 7.21:

$$N_e = \frac{N_i}{(1 + K_b t^*)^n} = N_i(8·75)^{-n}$$

Therefore for $N_i = 4 \times 10^7$ and $N_e = 100$, n is given by:

$$n = \frac{\log(N_i/N_e)}{\log(8·75)} = \frac{\log(4 \times 10^5)}{\log(8·75)} = \frac{5·602}{0·942}$$

$$= 6$$

Thus six ponds with $t^* = 5$ d will produce an effluent with an acceptable FC count. Now consider BOD_5 removal:
(1) The volumetric loading on the first pond is given by equation 7.9 as:

$$\lambda_v = L_i/t^* = 500/5 = 100 \text{ g/m}^3 \text{ d}$$

The first pond is therefore anaerobic (see section 7.8) and the BOD_5 removal in it may be estimated as $\frac{4}{5} \times (70\%) = 56\%$ (see Section 7.11). The effluent from the first pond therefore has a BOD_5 of $0·44 \times 500 = 220$ mg/l.
(2) The second pond is facultative and therefore its effluent BOD_5 can be estimated from equation 4.7 and 7.12 m:

$$L_e = \frac{L_i}{1 + [0·3(1·05)^{T-20}t^*]}$$

$$= \frac{220}{1 + (0·3 \times 1·05^{-3} \times 5)} = 96 \text{ mg/l}$$

(3) The four remaining (maturation) ponds are easily able to reduce the BOD_5 from 96 mg/l to < 25 mg/l.

8

Aerated Lagoons

8.1 OPERATION

Aerated lagoons are activated sludge units operated *without* sludge return. Historically they were developed from waste stabilization ponds in temperate climates where mechanical aeration was used to supplement the algal oxygen supply in winter. It was found, however, that soon after the aerators were put into operation the algae disappeared and the microbial flora resembled that of activated sludge. Aerated lagoons (Figure 8.1) are now usually designed as completely mixed non-return activated sludge units. Floating aerators (Figure 8.2) are most commonly used to supply the necessary oxygen and mixing power.

Aerated lagoons achieve BOD_5 removals > 90 per cent at comparatively long retention times (2–6 d); retention times < 2 d are not recommended as they are too short to permit the development of a healthy flocculent sludge (even so the activated sludge concentration is only 200–400 mg/l, in contrast to the 2000–6000 mg/l found in conventional systems and oxidation ditches). They are often useful as pretreatment units before a series of ponds, particularly when used as a second stage of development to extend the pond capacity. In common with all activated sludge systems, aerated lagoons are not particularly

Figure 8.1 An aerated lagoon [Courtesy of Peabody Welles]

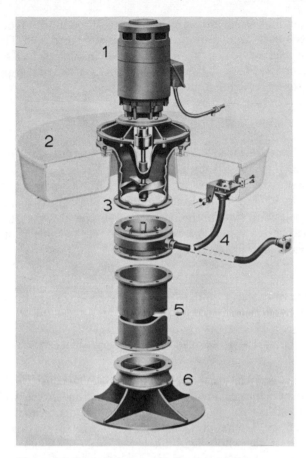

Figure 8.2 Exploded view of a floating surface aerator.
1, motor; 2, float; 3, propeller; 4, mooring lines; 5,
draught tube (optional); 6, intake plate. Water is sucked
up the draught tube and thrown out through the collar
immediately below the motor (see Figure 8.1) [Courtesy
of Peabody Welles]

effective in removing faecal bacteria: FC reductions are only 90–95 per cent and
further treatment may therefore be necessary (Section 11.3).

8.2 DESIGN

The rate of oxidation of a waste in an aerated lagoon has been found to be
well approximated by a first order equation, e.g. equation 4.7:

$$L_e = \frac{L_i}{1 + k_1 t^*} \tag{4.7}$$

Now L_e is the BOD_5 of the effluent which is due to two separate fractions:

(1) the small amount of the influent waste not oxidized in the lagoon and (2) the bacterial cells synthesized during oxidation (see Figure 3.2). These fractions are generally referred to as the 'soluble' and 'insoluble' BOD respectively. It is convenient (and in fact conceptually more correct) to apply first order kinetics only to the removal of the soluble fraction:

$$F_e = \frac{L_i}{1 + \kappa t^*} \tag{8.1}$$

where F_e = soluble BOD_5 in the effluent (i.e. the fraction of the influent BOD_5 which escapes oxidation; see Figure 3.2), mg/l

κ = first order rate constant for soluble BOD_5 removal, d^{-1}.

It should be noted that all the influent BOD_5 is assumed to be soluble (i.e. $F_i = L_i$). The retention time in the lagoon is 2–6 d, with 4 d as the most usual value. A typical design value for κ is 5 d^{-1} at 20 °C; its value at other temperatures can be estimated from the equation:

$$\kappa_T = 5(1 \cdot 035)^{T-20} \tag{8.2}$$

The quantity of bacteria synthesized in the lagoon is related to the quantity of soluble BOD_5 oxidized:

$$\frac{dX}{dt} = Y\frac{dF}{dt} \tag{8.3}$$

where X = cell concentration in lagoon, mg/l

Y = yield coefficient (defined by this equation as the weight of cells formed per unit weight of soluble BOD_5 consumed).

Y is typically 0·6–0·7. On a finite time basis, say one retention time, equation 8.3 can be rewritten for the whole lagoon as:

$$\frac{XV}{t^*} = \frac{Y(L_i - F_e)V}{t^*} \tag{8.4}$$

where V = lagoon volume, m^3.

The rate of cell synthesis must be balanced by the sum of the rates which cells leave the lagoon in the effluent and at which they die in the lagoon. The rate at which the cells leave the lagoon is QX where Q is the flow through the lagoon. The rate at which some of the cells in the lagoon die is proportional to the quantity of cells present; it is usually given as bXV where b is the rate of autolysis in d^{-1} (see equation 2.4; typically $b = 0·07$ d^{-1} at 20 °C). Thus:

$$\frac{Y(L_i - F_e)V}{t^*} = bXV + QX \tag{8.5}$$

$$\begin{pmatrix} \text{rate of} \\ \text{synthesis} \end{pmatrix} = \begin{pmatrix} \text{rate of} \\ \text{autolysis} \end{pmatrix} + \begin{pmatrix} \text{rate of loss} \\ \text{in effluent} \end{pmatrix}$$

Rearranging and writing V/Q as t^*:

$$X = \frac{Y(L_i - F_e)}{1 + bt^*} \tag{8.6}$$

This quantity of cells X can be converted to an equivalent ultimate BOD by considering the chemical equation for their complete oxidation:

$$C_5H_7NO_2 + 5\,O_2 \rightarrow 5\,CO_2 + 2\,H_2O + NH_3$$

$$\underset{\text{(cells)}}{113} \qquad 5 \times 32$$

Thus 1 g of cells has an ultimate BOD of $(5 \times 32/113) = 1.42$ g. Since $BOD_5/BOD_u = \frac{2}{3}$, 1 g of cells has a BOD_5 of 0.95 g. Thus the effluent BOD_5 L_e is given by:

$$L_e = F_e + 0.95X \tag{8.7}$$

Oxygen requirement

The quantity of oxygen required for bio-oxidation is the amount of total (i.e. soluble + insoluble) ultimate BOD removed:

$$R_{O_2} = 1.5(L_i - L_e)Q \tag{8.8}$$

Substituting equation 8.7:

$$R_{O_2} = 1.5(L_i - F_e)Q - 1.42XQ \tag{8.9}$$

i.e. the oxygen requirement is the ultimate soluble BOD removed less the ultimate BOD due to the cells wasted in the effluent.

8.3 AERATOR PERFORMANCE

The aerator must supply both sufficient oxygen for bio-oxidation and sufficient power to mix the lagoon contents. The power required for mixing ($5\ W/m^3$) is usually greater than that for oxygen supply.

Manufacturers certify that their aeration equipment has an *oxygen transfer rate* of so many kg O_2/kWh (or lb/h.p.-hour) under standard test conditions (these are: tap water as the test liquid, at 20 °C and initially with zero dissolved oxygen concentration). This standard rating has to be corrected for field conditions as follows:[1]

$$N = N_0[\alpha][(1.024)^{T-20}]\left[\frac{\beta c_{s(T,A)} - c_L}{c_{s(20,\,0)}}\right] \tag{8.10}$$

where N = oxygen transfer rate in field
N_0 = oxygen transfer rate under standard test conditions.

The first correction term is to allow for the nature of the liquid to be aerated:

α = ratio of the oxygen transfer rate in the waste to that in tap water at the same temperature (typically for domestic sewage, $\alpha = 0.7$).

The second term is an Arrhenius temperature correction term. The third term is the dissolved oxygen correction which allows for the difference between the DO concentration in the lagoon (typically 1–2 mg/l) and that adopted in the standard rating test (zero):

$c_{s(T,A)}$ = O_2 saturation concentration (solubility) in distilled water at temperature T and altitude A. The values of $c_{s(T)}$ at sea level (760 mm Hg) are given in Table 8.1 for $T = 15$–35 °C. The correction for altitude is made by considering the mean air pressure P_A (mm Hg) at that altitude:

$$c_{s(T,A)} = c_{s(T)}(P_A/760)$$

$c_{s(20,0)}$ = O_2 saturation concentration in distilled water at 20 °C and sea level (= 9·08 mg/l)

c_L = DO concentration in lagoon (1–2 mg/l)

β = ratio of O_2 saturation concentration in the waste to that in distilled water (typically for domestic sewage, $\beta = 0·9$).

Modern floating surface aerators are rated at $1·2 - 2·4$ kg O_2/kWh.

Table 8.1 Solubility of oxygen in distilled water at sea level (760 mm Hg) for various temperatures*

Temperature (°C)	Solubility (mg/l)
15	10·07
16	9·86
17	9·65
18	9·46
19	9·27
20	9·08
21	8·91
22	8·74
23	8·57
24	8·42
25	8·26
26	8·12
27	7·97
28	7·84
29	7·70
30	7·57

*From H. A. Montgomery et al., Journal of Applied Chemistry, **14**, 280 (1964).

8.4 CONSTRUCTION

The construction of aerated lagoons is essentially the same as that of waste stabilization ponds. The major differences are: greater depths (usually 3–5m), steeper embankment slopes (1 to 1·5–2) and frequently the provision of a complete butyl rubber or polythene lining to prevent scour by the turbulence induced by the aerators.

The location of the surface aerators is an important factor. If they are too close the overall oxygen transfer rate is reduced owing to eddy interference; for example 75 kW (100 h.p.) aerators require a minimum spacing of about 20 m.[2]

8.5 EFFLUENT TREATMENT

Consider a typical domestic waste with $L_i = 500$ mg/l. For $t^* = 4$ d and $T = 20$ °C and assuming $\kappa = 5$ d^{-1}, $b = 0.07$ d^{-1} and $Y = 0.65$, equations 8.1, 8.5 and 8.7 give:

$$F_e = \frac{L_i}{1 + kt^*} = \frac{500}{1 + (5 \times 4)} = 24 \text{ mg/l}$$

$$X = \frac{Y(L_i - F_e)}{1 + bt^*} = \frac{0.65(500 - 24)}{1 + (0.07 \times 4)} = 240 \text{ mg/l}$$

$$L_e = F_e + 0.95X = 24 + (0.95 \times 240) = 252 \text{ mg/l}$$

Thus the BOD$_5$ of the lagoon effluent is 252 mg/l, *but 90 per cent of this is due to the bacteria present*. If these cells (or most of them) are removed from the effluent prior to discharge, the effluent BOD$_5$ will be considerably reduced. There are two ways to do this: (1) discharge into a series of maturation ponds and (2) sedimentation with subsequent digestion of the settled sludge.

Discharge into maturation ponds

This method is the more advantageous as it also achieves a high degree of bacteriological purification. The first pond acts as a settling basin and, to allow for the accumulation of sludge, it should have a retention time of 10 d and a depth of 1.5–2.0 m. The remaining ponds have a retention time of 5 d and a depth of 1.0–1.5 m; their number depends on the required degree of faecal coliform removal (use equations 7.21 and 7.22 with the assumption of 90 per cent FC removal in the aerated lagoon).

Sedimentation and sludge digestion (Figure 8.3)

The lagoon effluent enters a conventional secondary sedimentation tank and the supernatant is discharged into the receiving watercourse or used for irrigation (chlorination prior to discharge or re-use may be advisable; see Section 11.3). In order to avoid odour problems the sludge requires further treatment before it can be put on the drying beds. It is pumped to an *aerobic digester*, which is a small aerated lagoon, where it is aerated for 4–10 d (the aeration requirements are those for complete mixing).[3] After this period of intense

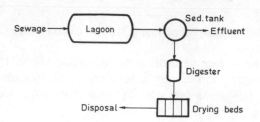

Figure 8.3 Flow diagram for aerated lagoon
system incorporating sludge digestion

aerobic stabilization the sludge is sufficiently mineralized to be placed on drying beds without fear of odour release.

8.6 REFERENCES

1. Nogaj, R. J., *Chemical Engineering, New York*, **79**, (4), 95 (1972).
2. Price, K. S. *et al.*, *Journal of the Environmental Engineering Division, American Society of Civil Engineers*, **99**, 283 (1973).
3. Cook, E. E. *et al.*, *Public Works, New York*, **102**, (11), 69–72 (1971).

Further reading

Balasha, E. and Sperber, H., 'Treatment of domestic wastes in an aerated lagoon and polishing pond', *Water Research*, **9**, 43 (1975).

8.7 DESIGN EXAMPLE

Design an aerated lagoon to treat 1000 m^3/d of domestic sewage; the BOD_5 is 400 mg/l and the required reduction is 85 per cent. The design temperature is 20 °C. What effluent treatment system would you recommend if the effluent is to be (a) discharged into an estuary and (b) used to irrigate a sisal plantation? The SS of the raw waste is 350 mg/l and the effluent SS is to be < 50 mg/l.

Solution

Take $t^* = 4$ d, $\kappa = 5$ d^{-1}, $b = 0\cdot07$ d^{-1} and $y = 0\cdot65$. Then from equations 8.1, 8.5 and 8.7:

$$F_e = \frac{L_i}{1 + \kappa t^*} = \frac{400}{1 + (5 \times 4)} = 19 \text{ mg/l}$$

$$X = \frac{Y(L_i - F_e)}{1 + bt^*} = \frac{0\cdot65(400 - 19)}{1 + (0\cdot07 \times 4)} = 194 \text{ mg/l}$$

$$L_e = F_e + 0\cdot95X = 19 + (0\cdot95 \times 194) = 203 \text{ mg/l}$$

Lagoon size: Assume a depth of 3 m. The lagoon mid-depth area is given by equation 7.6 as:

$$A = \frac{Qt^*}{D} = \frac{1000 \times 4}{3} = 1340 \text{ m}^2$$

Choose two lagoons in parallel, each 26 m square at mid-depth.

Aeration: Estimate the quantity of oxygen required from equation 8.8:

$$R_{O_2} = 1\cdot5(L_i - L_e)Q = 1\cdot5(400 - 203)1000$$

$$= 295\,000 \text{ g/d} = 12\cdot3 \text{ kg/h}$$

Assume that the aerators have a standard rating of 2 kg O_2/kWh and correct for field conditions from equation 8.10 (assume $\alpha = 0\cdot7$, $\beta = 0\cdot9$, $c_L = 1$ mg/l and zero altitude):

$$N = N_0\alpha(1\cdot024)^{T-20}\left(\frac{\beta c_{s(T,A)} - c_L}{c_{s(20,0)}}\right)$$

$$= 2 \times 0\cdot7 \times 1 \times \left(\frac{(0\cdot9 \times 9\cdot08) - 1}{9\cdot08}\right)$$

$$= 1\cdot1 \text{ kg } O_2/\text{kWh}$$

Aerator power required $= 12\cdot3/1\cdot1 = 11\cdot2$ kW.
Power for complete mixing (@ 5 W/m³) $= AD \times 5 = 1340 \times 3 \times 5 = 20\,000$ W.
Therefore choose one central 10 kW aerator for each lagoon.

 Effluent treatment: (a) discharge into an estuary: To avoid estuarine pollution, the solids must be removed from the lagoon effluent. The sedimentation tank is designed on the basis of an overflow rate of 22 m³/m² d and a retention time of 2 h at $3 \times DWF$. The overflow is the inflow less the underflow. Using the notation of Figure 8.4 we can apply a solids balance across the sedimentation tank:

$$Q_i c_i = Q_o c_o + Q_u c_u$$

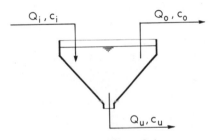

Figure 8.4

Now $Q_u = Q_i - Q_o$, so that this equation can be written as:

$$Q_o = Q_i\left(\frac{c_u - c_i}{c_u - c_o}\right)$$

Assuming that half the solids in the raw waste are non-biodegradable, c_i is given by:

$$c_i = X + 0\cdot5(\text{SS}) = 194 + 175 \doteqdot 370 \text{ mg/l}$$

Now $c_o = 50$ mg/l (the required effluent standard). Therefore, assuming that the solids are thickened to an underflow concentration of 5000 mg/l, Q_o is given as:

$$Q_o = Q_i\left(\frac{5000 - 370}{5000 - 50}\right)$$

$$= 0\cdot95\, Q_i$$

$$= 0\cdot95 \times 3 \times 1000 = 2850 \text{ m}^3/\text{d}$$

Therefore the dimensions of the tank are:

$$\text{Surface area} = 2850/22 = 130 \text{ m}^2 \text{ (12·9 m dia.)}$$

$$\text{Depth} = \frac{2850 \times (2/24)}{130} = 1\cdot83 \text{ m}$$

The underflow (sludge flow) is:

$$Q_u = Q_i - Q_o = 3000 - 2850 = 150 \text{ m}^3/\text{d}$$

The volume V of the aerobic digester ($t^* = 10$ d) is given by:

$$V = Qt^* = 150 \times 10 = 1500 \text{ m}^3 \text{ (say, } 22\cdot5 \times 22\cdot5 \times 3 \text{ m at mid-depth)}$$

The aeration requirement is that for complete mixing: $1500 \times 5 = 7500$ W; provide one central $7\cdot5$ kW aerator.

The drying bed area is calculated on the assumption that the sludge takes 5 d to dry when poured to a depth of $0\cdot5$ m (this depth is greater than the $0\cdot25$ m normally used for conventional drying beds because the solids concentration is lower) Therefore:

$$\text{area} = \frac{\text{sludge flow (m}^3/\text{d)} \times \text{drying time (d)}}{\text{depth (m)}}$$

$$= \frac{150 \times 5}{0\cdot5} = 1500 \text{ m}^2$$

Effluent treatment: (b) irrigation of sisal plantation: A single settling pond would suffice in this case as there is no need to attempt any further FC reduction in the effluent as it is to be used to irrigate an industrial crop. Assume $t^* = 10$ d and $D = 2$ m. From equation 7.6:

$$A = Qt^*/D = 1000 \times 10/2 = 5000 \text{ m}^2$$

Choose two ponds in parallel, each 26×96 m at mid-depth. The effluent SS may be expected to be about 50–100 mg/l (this is not unduly high for irrigation water; the settling pond will however achieve a substantial removal of parasite eggs).

9

Oxidation Ditches

9.1 OPERATION

The oxidation (Pasveer) ditch is a modification of the conventional activated sludge process.[1] Its essential operational features are that it receives screened or comminuted raw sewage and provides long retention times: the hydraulic retention time is commonly 0·5–1·5 d and that for the solids 20–30 d. The

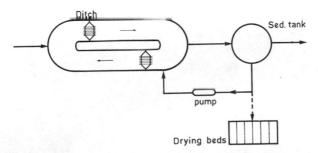

Figure 9.1 Flow diagram for oxidation ditch

Figure 9.2 Typical oxidation ditch installation [Courtesy of Lakeside Equipment Corporation]

Figure 9.3 Cage rotors [Courtesy of Whitehead & Poole Ltd]

Figure 9.4 Mammoth rotor [Courtesy of Whitehead and Poole Ltd]

latter, achieved by recycling > 95 per cent of the sludge, ensures minimal excess sludge production and a high degree of mineralization in the sludge that is produced. Sludge handling and treatment is almost negligible since the small amounts of waste sludge can be readily dewatered without odour on drying beds. The two other major differences from the conventional process lie in shape and type of aerator. The oxidation ditch is a long continuous channel, usually oval in plan and 1·0–1·5 m deep (Figures 9.1 and 9.2). The ditch liquor is aerated by one or more cage rotors (Figure 9.3) placed across the channel; for large flows it is usually more economic to provide the more powerful mammoth rotors[2] (Figure 9.4). The rotors also impart a velocity of 0·3–0·4 m/s to the ditch contents, sufficient to maintain the active solids in suspension.

The concentration of total SS in the ditch is 3000–5000 mg/l; in order to prevent the concentration much exceeding this range the return sludge flow is diverted to the drying beds for a short period each day; this period is best determined by operational experience (a simple field check on the ditch SS concentration is to fill a 1000 ml graduated cylinder to the mark with the ditch liquor; if the solids concentration is 3500–4500 mg/l the volume of sludge which settles in 30 minutes should be about 200 ml). Alternatively the sludge wastage rate may be estimated by considering the solids retention time (Section 9.2). BOD_5 removals are consistently > 95 per cent.

The oxidation ditch was developed in Holland to provide small communities of 200–15 000 people with sewage treatment facilities at the same per capita cost as conventional works serving much larger populations. At present there are few oxidation ditches in hot climates since waste stabilization ponds are usually more favourable both in terms of cost and the removal of faecal bacteria although where there is a reliable electricity supply but insufficient land for pond they are being increasingly used (but unfortunately seldom with provision for improving the bacteriological quality of the effluent).

(In Europe for populations < 1000 the ditch usually serves the dual purpose of aeration tank and sedimentation tank; one to three times each day the sewage inlet is closed and the rotor switched off for a period of about 1 h to permit the bioflocs to settle and so produce a high quality supernatant (during this time the influent sewage is stored in the sewer). After settlement some of the supernatant is discharged to the receiving watercourse; the inlet is then opened and the aerator switched on. Excess sludge is accumulated in a sludge trap adjacent to the ditch and is removed at regular intervals (normally once a day) and spread on drying beds. For populations > 1000 it is more economic to have continuous operation by providing a sedimentation tank and a sludge return pump. In tropical developing countries oxidation ditches are unlikely to be used for populations < 1000 for which waste stabilization ponds are usually more suitable. The type of ditch shown in Figure 9.1 can therefore be expected to be the one most frequently used in hot climates.)

9.2 DESIGN

Designs are purely empirical at the present time. The depth is in the range 1–2 m and the volume is dependent on the retention time which in turn is based on the sludge loading factor γ. This is the weight of BOD_5 applied to the ditch liquor suspended solids per day; it is measured in g BOD_5 per g solids per d (units: d^{-1}). The weight of BOD_5 entering the ditch is L_iQ g/d, where L_i is the influent BOD_5 (mg/l, $= g/m^3$) and Q the flow (m^3/d); the weight of suspended solids in the ditch is SV where S is the ditch liquor suspended solids concentration (mg/l) and V the ditch volume (m^3). Therefore the sludge loading factor is:

$$\gamma = \frac{L_iQ}{SV} \tag{9.1}$$

or, since $V/Q = t^*$ (the mean hydraulic retention time, d):

$$\gamma = \frac{L_i}{St^*} \tag{9.2}$$

Table 9.1 Design criteria for oxidation ditches in
Europe and India*

Parameter	India	Europe
Sludge loading factor (d^{-1})	0·1–0·3	0·05
Aeration requirement (kgO_2/kg BOD_5 applied)	1·5–2·0	2·0
Excess sludge production (g/hd d)	5–10	25–30
Area of sludge drying beds (m^2/hd)	0·025	0·35
Overall land requirement (m^2/hd)	0·125	1·2

*From Arceivala and Alagarsamy.[3]

Design values commonly used in Europe are $\gamma = 0·05$ d^{-1} and $S = 4000$ mg/l
which give for a typical domestic waste ($L_i = 300$ mg/l) a retention time of
1·5 d. In the tropics, however, much higher loadings (and therefore shorter
retention times) are possible:[3] a comparison between the design criteria used
in temperate and hot climates is given in Table 9.1.

Aeration

Oxygen is required at a rate of 1·5–2·0 g O_2/g BOD_5 applied.[3,4] Such a rate
of supply includes an allowance for the endogenous respiration of the sludge
and maintains aerobic conditions along the entire length of the ditch.

Cage rotors (70 cm dia.) have a standard rating of 3 kg O_2/m length per h
at the usual immersion depth of 150 mm and when rotating at 70 rpm. Mammoth
rotors (1 m dia.) are rated at 10 kg O_2/m h at 300 mm immersion and 70 rpm.

Sludge wastage

To control the rate of solids accumulation in the ditch, a proportion of the
ditch solids must be wasted each day. The rate of wastage is governed by the
desired solids retention time (SRT): thus if the desired SRT is t_s^* d then $100/t_s^*$
per cent of the ditch solids must be wasted each day. This can be achieved by
diverting $100/t_s^*$ per cent of the flow from the ditch directly to waste. However,
it is more usual to waste the sludge from the sludge return line. The percentage
$100/t_s^*$ must therefore be corrected for the change in concentration due to
sedimentation. Noting that the inflow concentration to the sedimentation tank
is the ditch SS concentration S, the quantity of waste sludge to be diverted to
the drying beds, *expressed as a percentage of the raw waste flow,Q, is*

$$\frac{100}{t_s^*}\left(\frac{S}{S_R}\right)$$

where S_R = the solids concentration in the sludge return line (= underflow
concentration from sedimentation tank).

For example, choosing $t_s^* = 30$ d and $S = 4000$ mg/l and assuming that $S_R = 50\,000$ mg/l, the sludge wastage rate would be 0·27 per cent. This illustrates the extremely low rate of sludge production in oxidation ditches.

9.3 CONSTRUCTION

Preferably the ditch should have a concrete lining with side slopes of about 1 in $1\frac{1}{2}$. A lining of butyl rubber is satisfactory, though of course not as long lasting as one of concrete. If a flexible lining is used, a rigid lining of concrete or concrete slabs should be placed under the rotors and for a distance 5 m downstream; this is to prevent damage due to the high turbulence in these areas. When mammoth rotors are used, the ditch depth is 3–5 m; the walls are vertical and constructed from reinforced concrete.

9.4 CAROUSEL DITCH

The carousel ditch is a development[5] of the oxidation ditch in which oxygenation of the waste is achieved by fixed surface aerators rather than by cage rotors. The surface aerators (Figure 6.9) are normally placed at one end of the ditch (Figure 9.5) and the liquid depth is 2–4 m. The design and operation are otherwise the same as for the oxidation ditch. Fewer aerators and greater

Figure 9.5 Typical carousel ditch installation [Courtesy of Simon Hartley Ltd]

economy of land, particularly at high flows, are the advantages claimed for the carousel ditch (however these are also achieved by oxidation ditches which incorporate mammoth rotors).

9.5 REFERENCES

1. Baars, J. K., *Bulletin of the World Health Organization*, **26**, 465 (1962).
2. Rees, J. T. and Skellet, C. F., *Water Pollution Control*, **73**, 608 (1974).
3. Arceivala, S. J. and Alagarsamy, S. R., in *Low Cost Waste Treatment*, Central Public Health Engineering Research Laboratory, Nagpur, 1970.
4. *Technical Memorandum on Activated-Sludge Sewage Treatment Installations providing for a long period of Aeration*, HMSO, London, 1969.
5. Koot, A. C. J. and Zepper, J., *Water Research*, **6**, 401 (1972).

9.6 DESIGN EXAMPLE

Design an oxidation ditch scheme to serve a suburban area which has a population of 10 000. The effluent flow is 80 l/hd d and the BOD_5 contribution may be taken as 40 g/hd d. The design temperature is 20 °C and the ditch site is at sea level.
Solution

Flow $Q = 80 \times 10\ 000$ l/d $= 800$ m^3/d
Influent BOD$_5$ $L_i = 40 \times 1000/80 = 500$ mg/l
From Table 9.1 choose $\gamma = 0.20$ d^{-1} and $S = 400$ mg/l
From equation 9.1:

$$V = \frac{L_iQ}{S} = \frac{500 \times 800}{4000 \times 0.20} = 500 \text{ m}^3$$
$$t^* = V/Q = 500/800 \text{ d} = 15 \text{ h}$$

The oxygen requirement is taken as twice the BOD$_5$ load:

$$R_{O_2} = (2 \times 40 \text{ g/hd d}) \times (10\ 000 \text{ people})$$
$$= 800 \text{ kg/d} = 34 \text{ kg/h}$$

The cage rotor rating is now corrected for field conditions using equation 8.10:

$$N = N_0\alpha(1.024)^{T-20} \left(\frac{\beta c_{s(T,A)} - c_L}{c_{s(20,0)}} \right)$$
$$= 3 \times 0.7 \times 1 \times \left(\frac{(0.9 \times 9.08) - 1}{9.08} \right)$$
$$= 1.64 \text{ kg O}_2/\text{m h}$$

The required length of rotor is $34/1.64 = 20.7$ m.

Sedimentation tank: The area is based on a peak overflow rate of 22 m^3/m^2 d and a retention time of 2 h at peak flow. Since the sludge flow is returned to the ditch (ignoring the small volume wasted), the peak overflow is taken as 3 × inflow (a solids mass balance as used for the design of aerated lagoons in Section 8.7 is therefore unnecessary). Therefore:

$$\text{area of tank} = (3 \times 800)/22 = 109 \text{ m}^2 \ (11.8 \text{ m dia})$$
$$\text{depth} = (3 \times 800) \times (2/24)/109 = 1.83 \text{ m}$$

Drying beds: Assume an area of 0.025 m^2/hd (Table 9.1)
Therefore:

$$\text{area} = 0.025 \times 10\ 000 = 250 \text{ m}^2.$$

10

High-Rate Biofiltration

10.1 PROCESS DESCRIPTION

High-rate biofiltration is a development of the standard (low-rate) biological filter used in conventional secondary treatment (Section 6.3). High-rate bio-filters contain a plastic packing, usually made from high-density PVC, of open structure and regular geometry (Figure 10.1) which serves as the support medium for the microbial film. The voids ratio is > 90 per cent and this effectively prevents blockage by excess microbial growth or other solids and also ensures an adequate flow of air through the filter so that the bacteria are supplied with sufficient oxygen. The low density and high mechanical strength of plastic media minimizes structural requirements (Figure 10.2).

High-rate biofilters are sometimes used in temperate climates to treat domestic sewage (usually to relieve an overloaded conventional works) but

Figure 10.1 Module of high-rate plastic filter medium, 'Flocor'
[Courtesy of ICI Pollution Control Systems]

Figure 10.2 Biofiltration tower treating dairy wastes
[Courtesy of ICI Pollution Control Systems]

they are most commonly employed for the treatment of biodegradable industrial wastes. In hot climates it is probable that their major function will be the treatment of industrial wastes, particularly those from factories in the crowded industrial areas of major towns where space for other types of treatment plant is unavailable.

10.2 PARTIAL TREATMENT OF STRONG WASTES

High-rate biofilters have a large capacity for BOD removal at high load: removals of up to about 7–8 kg BOD_5/m^3d can be obtained in filters receiving comminuted sewage. Although it is possible to design a series of biofilters to provide an effluent of $BOD_5 < 25$ mg/l, it is not generally economic to do so as the volume of plastic medium required becomes too expensive. High-rate biofiltration is most advantageously used for the *partial* treatment of domestic and industrial wastes, particularly those of high strength, and when pretreatment is required by the local authority prior to discharge into its sewers. For this purpose they have been extensively studied at the Water Pollution Research Laboratory, Stevenage, England:

111

In an investigation in which a simulated milk-processing waste water was given partial treatment on a pilot biological filter containing a plastic medium, the BOD_5 of the feed was varied over the range 500–1500 mg/l and the performance was studied at three different hydraulic loadings. The results [Figure 10.3] indicate that first-order kinetics applied over the range studied and broadly confirm that the percentage BOD_5 removal obtained by biological filtration, at a given hydraulic loading, remains fairly constant over a wide range of feed strengths. [Figure 10.4] shows the relation between the weight of BOD_5

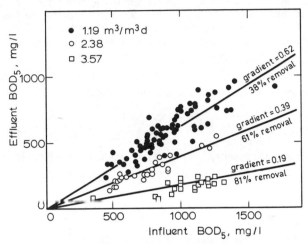

Figure 10.3 Variation of BOD_5 removal efficiency with hydraulic loading [From Bruce and Boon[1]]

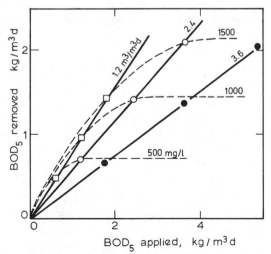

Figure 10.4 Effect of hydraulic loading (solid lines) and influent BOD_5 (broken lines) on BOD removal at various loadings, based on average results of Figure 10.3 [From Bruce and Boon[1]]

applied and the weight of BOD_5 removed, per unit volume of medium, over the range of operating conditions. It is seen that BOD_5 removal per unit volume of medium increased with load applied, as would be expected, but, for any given load applied, the actual load of BOD_5 removed varied according to the actual hydraulic loading and BOD_5 of the feed. The maximum amount removed per unit volume of medium occurred with low hydraulic load and high BOD_5 feed. This indicates that the real efficiency of the filter increased with strength of feed, since the BOD_5 of the effluent was about the same in each case. Thus it is an advantage to conserve the strength of the waste water so as to operate the lowest hydraulic loading possible rather than to dilute the waste water and to operate at higher hydraulic loads and weaker feed strengths, even though the BOD_5 load on the filter might be the same in each case. With increasing strength the driving force for BOD_5 removal by the film increases.[1]

At hydraulic loadings < 24 m^3/m^3 d recirculation is usually necessary to ensure that the whole cross-section is wetted by the flow.

10.3 DESIGN

High-rate biofilters are essentially plug flow reactors in which BOD_5 removal is reasonably well represented by first order kinetics. Equation 4.9 thus applies:

$$\frac{L_e}{L_i} = \exp(-k_1 t^*) \tag{4.9}$$

The retention time t^* is a function of the specific surface area S of the filter medium (its surface area per unit volume, m^2/m^3), the gross filter volume V (m^3) and the flow rate Q (m^3/d). Now the retention time in a biofilter is the ratio distance/velocity. The distance travelled is a zigzag path (see Figure 10.1) which is proportional to the height H of the biofilter, say αH (where $\alpha > 1$). To calculate the velocity ($=$ flow/unit area), consider a 1 m cube of the filter medium:

(1) The flow applied to the cube is Q/A, where A is the cross-sectional area of the biofilter.
(2) The cross-sectional area available for effective flow is the area immediately adjacent to the film of microbial growth. In the 1 m cube there are S m^2 of surface area. Since the depth of the cube is 1 m, its effective linear cross-sectional dimension ('flow perimeter') is S m (Figure 10.5). If the flow thickness is d, the cross-sectional area used for the flow is xSd where x is < 1 and is a combined measure of the non-availability and non-use of the theoretical flow area (Sd) for actual flow (non-availability arises from the presence of the microbial film and non-use is due to the non-ideal distribution of the waste over the whole cross-section).

The flow velocity is (flow/area), $= (Q/A)/(xSd)$. The retention time is (distance/velocity), so that:

$$t^* = \frac{\alpha H}{(Q/A)/(xSd)} \tag{10.1}$$

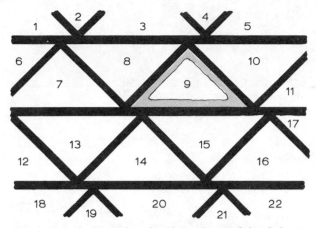

Figure 10.5 Horizontal section through a module of plastic
filter medium (compare Figure 10.1). The 'flow perimeter'
per unit area is the sum of the perimeters of all the triangles
(such as those numbered 1–22) contained in 1 m². The
presence of the microbial film, as indicated in triangle 9,
reduces the flow perimeter from the value so calculated to
its real value

Equation 4.9 is thus rewritten as:

$$\frac{L_e}{L_i} = \exp -\frac{KVS}{Q} \qquad (10.2)$$

where $K = k_1 \alpha x d$ (units: m/d)

$V = AH$

The ratio Q/V is the volumetric hydraulic loading, m^3/m^3d. Since k_1 is temperature dependent (equation 4.8), the modified rate constant K also varies with
temperature. Pilot plant studies in England[2] have indicated that, using a
reference temperature of 15 °C, the equation for K_T is:

$$K_T = 0.037(1.08)^{T-15} \qquad (10.3)$$

Equation 10.3 should be considered only a first approximation at temperatures
> 25 °C.

10.4 REFERENCES

1. Bruce, A. M. and Boon, A. G., *Water Pollution Control*, **70**, 487 (1970).
2. *Water Pollution Research 1972*, HMSO, London, 1973, p. 45.

Further reading

Chipperfield, P. N. J. *et al.*, 'Multiple-stage, plastic-media treatment plants', *Journal of
the Water Pollution Control Federation*, **44**, 1955–1967 (1972).

10.5 DESIGN EXAMPLE

Design a biofilter to remove half the BOD_5 from 400 m^3/d of a cannery waste which has a BOD_5 of 1200 mg/l. The design temperature is 17 °C.

Solution
Use, for example, 'Flocor' medium (Figure 10.1) which has $S = 85$ m²/m³ and is available in modules of 1·2 m × 0·6 m × 0·6 m. From equation 10.3:

$$K_T = 0.037 (1.08)^2$$

$$= 0.043 \text{ m/d}$$

Rearrangement of equation 9.2 gives:

$$V = -\frac{Q \ln(L_e/L_i)}{K_T S}$$

$$= -\frac{400 \times (-0.69)}{0.043 \times 85}$$

$$= 76 \text{ m}^3$$

Suitable dimensions are 3·6 m × 4·2 m in plan and 5·4 m high (i.e. 9 layers of 21 modules); actual volume = 81·6 m³. The BOD_5 loading is 5·9 kg/m³ d and the hydraulic loadings are 4·9 m³/m³d and 26 m³/m²d.

11

Tertiary Treatment

11.1 PURPOSE

Tertiary treatment is the collective name given to those processes which are used to improve the quality of an effluent from a conventional sewage treatment works: that is, an effluent which has received primary (sedimentation) and secondary (biological) treatment. By extension the term is used here for any process which is used to improve the effluent produced by the treatment processes described in Chapters 7–10.

Tertiary treatment processes are most commonly used to achieve additional removal of suspended solids (including algae) and a further reduction in faecal coliform numbers. SS removal also produces a substantial reduction in BOD_5 since a large proportion of the BOD_5 of an effluent is due to its SS—algae in maturation pond effluent, for example, account for about two-thirds of the effluent BOD_5. Algae are 50 per cent protein and their recovery from pond effluents can make a useful contribution to a community's food supply (see Section 13.2).

11.2 REMOVAL OF SUSPENDED SOLIDS

There are several tertiary treatment methods currently used to remove SS. The more important of these are flocculation and sedimentation or flotation, microstraining, sand filtration and clarification in a pebble bed. These processes are briefly described below.

Flocculation

Flocculation with alum $[Al_2(SO_4)_3.16 H_2O]$ followed by sedimentation is an effective method of removing algae from pond effluents. For example at Windhoek, Namibia, an alum dose of 110 mg/l and a settling time of $3\frac{1}{2}$ h achieved an 80 per cent reduction in SS and 65 per cent in BOD_5.[1]

Flotation rather than sedimentation, is used at Stockton, California, where 200 000 m^3/d of pond effluent are treated for algae removal.[2] Flotation is an alternative method to sedimentation for solid–liquid separation: air is introduced to the flotation tank via fine diffusers and the minute air bubbles so formed attach themselves to the solid particles which then have sufficient bouyancy to rise to the surface where they float until removed by a mechanical skimming device. The advantages of flotation are that a shorter retention time is possible ($< \frac{1}{2}$ h) and that the resulting sludge is more concentrated. However these advantages are usually only significant when the initial SS concentration is high (> 150 mg/l).

Microstraining

Microstrainers (Figure 11.1) were originally developed for water treatment but they have been successfully used to reduce wastewater effluent SS from about 25–120 mg/l to < 20 mg/l.[3,4]

The microstrainer consists of a drum revolving on a horizontal axis at a peripheral speed of up to 0·5 m/s and having the curved surface covered with a stainless-steel fabric of special weave. [The fabric apertures are 60, 35 or, most commonly, 23 μm.] The effluent to be treated enters the drum through one end and passes out through the fabric, the other end of the drum being closed. Solids retained by the fabric are continually washed off by strained effluent sprayed under pressure through a row of jets mounted above the drum. Wash water containing the removed solid matter is collected in a trough inside the drum and is returned to the primary sedimentation tanks; in volume it may amount to about 5 per cent of the volume strained. A high-intensity ultra-violet lamp is mounted alongside the wash-water jets to prevent the formation of biological slimes on the fabric. The rate of flow through the microstrainer is determined by the head applied (which is limited to a maximum safe value of about 150 mm by a weir system), and by the concentration and nature of the suspended matter in the effluent.[5]

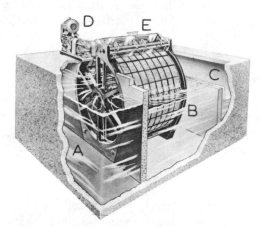

Figure 11.1 Microstrainer. A, dirty water inlet; B, microstrainer drum; C, clear water outlet, D, drive motor; E, backwash spray nozzles. The suspended matter is held against the inside of the drum from where it is washed away by the backwash spray nozzles; it then falls into a waste hopper connected to a waste outlet pipe concentric with the drum axis [Courtesy of Glenfield-Kennedy Ltd]

Microstrainers are extremely compact and are thus suitable for large works. They are commonly used to treat humus tank effluents and they are also suitable for aerated lagoon and oxidation ditch effluents. There is no reason in theory why they should not be used to treat maturation pond effluents, but in practice success has been limited.

Sand filtration

Rapid sand filters are best suited to large works where skilled maintenance is available.

Essentially the filters are similar to those employed at water works and consist of a bed of 1–2·5 mm sand 1–1·5 m deep supported on an under-drainage system provided with facilities for back-washing and air scouring. The liquid to be treated flows downwards through the sand at a controlled rate of about 120–240 m³/m²d. As filtration proceeds the loss of head increases because of the accumulation of suspended matter, and at a value of $2\frac{1}{2}$–3 m (or, more conveniently, at a certain time each day) the filter is back-washed using filtered effluent, air scour being employed to assist the separation of sludge from sand. The back-washings, amounting in volume to some $2\frac{1}{2}$ per cent of the volume filtered are returned to the inlet to the works.[5]

Reductions in SS range from 70 to 70 per cent with an associated BOD_5 removal of 50 per cent. However, at small works slow sand filters are more suitable:

These are usually 300–750 mm deep and consist of a layer of sand or fine pan ash, resting on a layer of coarse material which in turn rests on a system of drainage pipes. Construction is much simpler than that of the rapid sand filter since no back-washing facilities are provided. The effluent is allowed to percolate through the filter until the head loss becomes excessive. The filter is then drained and partially dried, and the surface layer of sludge is removed manually. Sand or pan ash is added from time to time to make good losses.[5]

When operated at a hydraulic loading of 2–3 m³/m²d, a slow sand filter can achieve SS reductions of 50–70 per cent.

Pebble bed clarifier[6]

This is an upward flow clarifier in which the effluent passes through a 150 mm layer of pea gravel (6–9 mm grading) which is supported by steel mesh (Figure 11.2). At a hydraulic loading of 20–25 m³/m²d, SS removal is about 50 per cent.

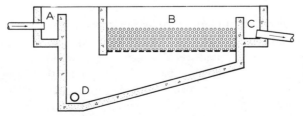

Figure 11.2 Banks filter. A, inlet weir; B, pebble bed on steel mesh support; C, effluent weir; D, sludge drain

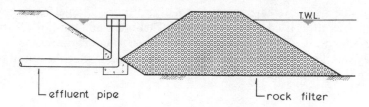

effluent pipe   rock filter

Figure 11.3 Horizontal rock filter for algae removal from waste
stabilization pond effluent

The solids which accumulate in the bed are flushed out by a jet of water (or effluent) when the water level has been reduced to below the bed.

An extension of the pebble bed clarifier is the *horizontal rock filter* (35–50 mm grading) proposed[7] for removing algae from stabilization pond effluent as it leaves the pond (Figure 11.3). Approximately 24 h retention time in the filter should be provided; assuming a voids ratio of 45 per cent, a filter volume of $0.18 \, m^3/hd$ is required for a waste flow of 80 l/hd d. The rock filter is therefore best suited to small works.

11.3 REMOVAL OF FAECAL BACTERIA

Additional removal of faecal bacteria can be achieved in maturation ponds or by disinfection. The number of faecal coliforms are used to indicate the bacteriological quality of the final effluent and the efficiency of the removal process.

Maturation ponds

These are most suitable for the treatment of effluent from aerated lagoons and conventional works although they can also be used after oxidation ditches and high-rate biofilters. The ponds should each have a retention time of 5–7 d and they should be designed in accordance with equations 7.21 and 7.22 (take $N_i = 4 \times 10^6$ FC/100 ml as aerated lagoons etc. achieve a 90–95 per cent reduction of FC).

Disinfection

Chlorine is the disinfectant in most common use. The technology of wastewater chlorination is similar to that used for drinking waters. The main difference is that, due to the large quantity of organic compounds present in sewage effluents, much greater chlorine doses are required, 10–25 mg/l not being uncommon. Contact times are also higher, 1–2 h as against 30 minutes. For best performance the chlorination tank should be designed to allow initial rapid mixing of the effluent with the chlorine feed followed by plug flow hydraulics to prevent short-circuiting. The quantity of chlorine needed to achieve the required FC reduction varies considerably from one effluent to another; it is best determined by experiment in each case. The chlorine content of the effluent should be continuously monitored as it leaves the tank. To avoid toxicity in the receiving stream it should never be $> 0\cdot1$ mg/l;[8] to achieve this, dechlorination with sulphur dioxide may be necessary, particularly if the effluent is being discharged into an area of biological importance.

11.4 REFERENCES

1. Van Vuuren, L. R. J. and van Duuren, F. A., *Journal of the Water Pollution Control Federation*, **37**, 1256 (1965).
2. Parker, D. S., *Water & Wastes Engineering*, **10** (1), 26 (1973).
3. Isaac, P. C. G. and Hibberd, R. L., *Water Research*, **6**, 465 (1972).
4. Diaper, E. W. J., *Water & Sewage Works*, **120** (8), 42 (1973).
5. *'Polishing' of Sewage Works Effluents (Notes on Water Pollution No. 22)*, HMSO, London, 1963.
6. Banks, D. H., *Surveyor*, 14 March 1964 and 16 January 1965.
7. O'Brien, W. J. *et al.*, *Water & Sewage Works*, **120** (3), 66 (1973).
8. Collins, H. F. and Deaner, D. G., *Journal of the Environmental Engineering Division, American Society of Civil Engineers*, **99**, 761 (1973).

12

Septic Tanks

12.1 OPERATION AND USE

Septic tanks are small, rectangular chambers, usually sited just below ground level, in which sewage is retained for 1–3 d. During this time the solids settle to the bottom of the tank where they are digested anaerobically. A thick crust of scum is formed at the surface and this helps to maintain anaerobic conditions. Although digestion of the settled solids is reasonably good some sludge accumulates and the tank must be desludged at regular intervals, usually once every 1–5 years. The effluent from septic tanks is, from a public health point of view, as dangerous as raw sewage and so requires further treatment before disposal. Although septic tanks are most commonly used to treat the sewage from individual households, they can be used as a communal facility for populations up to about 300. All the household wastewater should be led to the septic tank. In some older installations sullage is discharged directly to soakaways or open seepage channels; this is no longer recommended practice.

12.2 DESIGN

A two-compartment septic tank (Figure 12.1) is now generally preferred to

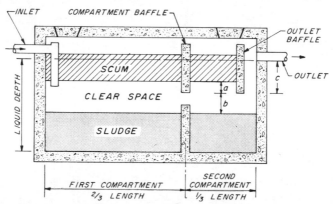

Figure 12.1 Two compartment septic tank: a, scum clear space (75 mm minimum); b, sludge clear space (300 mm minimum); $c = 40$ per cent of liquid depth [From Cotteral and Norris[3]]

one with only a single compartment as the suspended solids concentration in its effluent is considerably lower. The first compartment is usually twice the size of the second. The liquid depth is 1–2 m and the overall length to breadth ratio 2–3 to 1.

Experience has shown that in order to provide sufficiently quiescent conditions for effective sedimentation of the sewage solids, the liquid retention time should be at least 24 h. Two-thirds of the tank volume is normally reserved for the storage of accumulated sludge and scum, so that the size of the septic tank should be based on 3 d retention at start-up; this ensures that there is at least 1 d retention just prior to each desludging operation.

Expected desludging frequency

The tank should be emptied when it is approximately one-third full of sludge. The desludging interval (years) is therefore given by:

$$\frac{\frac{1}{3}\,(\text{tank volume, m}^3)}{(\text{sludge accumulation, m}^3/\text{hd yr}) \times (\text{population})}$$

Since the tank volume is given by:

$$(\text{waste flow, m}^3/\text{hd d}) \times (\text{population}) \times (3\text{ days' retention})$$

the desludging interval can be estimated from the ratio:

$$\frac{\text{waste flow (m}^3/\text{hd d)}}{\text{sludge accumulation (m}^3/\text{hd yr)}}$$

The rate of sludge accumulation is temperature dependent but, as is the case with anaerobic ponds (Section 7.4), the available field data are too few to establish a relationship between these two variables. The rate of accumulation of sludge was measured as 0·03–0·04 m^3/hd yr in Zambia[1] and the average value in South Africa was found to be 0·032 m^3/hd yr.[2] A value of 0·04 m^3/hd yr is probably a reasonable value for design.

Table 12.1 Setback requirements for septic tanks and drainfields*

Distance from	Septic tanks	Drainfields
	(m)	(m)
Buildings	1·5	3
Property boundaries	1·5	1·5
Wells	30	30
Streams	7·5[+]	30[+]
Cuts or embankments	7·5[+]	30[+]
Pools	3	7·5
Water pipes	3	3
Paths	1·5	1·5
Large trees	3	3

*From Cotteral and Norris.[3]
[+]These distances should be increased to 60 m if the installation is on a water supply watershed.

Location

The tank and its associated effluent disposal system must be sited away from wells and streams (to avoid their pollution) and from embankments and cuts (where the effluent might appear at the ground surface). Minimum distances are given in Table 12.1. It is particularly important to avoid the pollution of groundwaters used as a source of domestic supply.

12.3 DISPOSAL OF TANK EFFLUENT

Subsurface irrigation in drainfield trenches ('soakaways') is the most common method of disposal of septic tank effluent. The drainfield soil must of course be permeable; in impermeable soils either evapotranspiration beds or upflow filter must be used. For large flows (from populations above about 100) waste stabilization ponds may be more suitable; controlled flood irrigation may occasionally be an economic alternative.

Subsurface irrigation

The tank effluent is discharged directly to a number of drainage trenches connected in series with each other (Figure 12.2). Each trench consists of open-joint agricultural drainage tiles of 100 mm diameter laid on a 1 m depth of rock

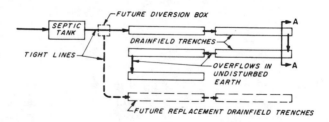

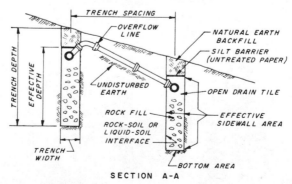

Figure 12.2 Drainage trenches for septic tank effluent
[From Cotteral and Norris[3]]

fill (20–50 mm grading). The effluent infiltrates into the soil surrounding the trench which eventually becomes clogged with sewage solids (provision must be therefore made to set aside land for use as a future replacement drainfield). Unless a more accurate figure is known, the rate of infiltration per m^2 of sidewall area may be conservatively estimated as 10 l/d.

Percolation tests

The soil must have a sufficient percolative capacity. This is determined by conducting percolation tests. A satisfactory field procedure[3] is to drill at least three 150 mm diameter test holes 0·5 m deep across the proposed drainfield. These are filled with water and left overnight so that the soil becomes saturated; on the following day, they are filled to a depth of 300 mm. After 30 and 90 minutes the water levels are measured; the soil is considered to have sufficient percolative capacity if the level in each hole has dropped 15 mm in this period of 1 h.

Evapotranspiration beds[2]

The effluent is distributed in open-joint pipes below the evapotranspiration bed which comprises a 200–500 mm depth of coarse sand and gravel underlying a 100 mm depth of topsoil planted with a fast-growing local grass. Grasses have high transpiration rates and the water content of the effluent is lost to the atmosphere by transpiration; the organic fraction is converted to grass which is periodically cut. In order to protect the bed from floods during the rainy seasons a bund should be constructed around the bed and suitable provision for surface water drainage made.

The size of the evapotranspiration beds is calculated on the basis of the transpiration rate being 80 per cent of the rate of evaporation from a free water surface, or on the basis of providing about 15 d storage (during the rainy seasons) of the effluent in the sand layer, whichever gives the larger area.

Upflow filters[4]

In an upflow filter the effluent enters at the base, flows upwards through a layer of coarse aggregate about 0·5 deep and is discharged over a weir at the top. Anaerobic bacteria grow on the surface of the aggregate and oxidize the effluent as it passes by. The head loss is low, about 30–150 mm during normal operation. Field studies in India have shown that these filters can effect a 70 per cent reduction in BOD and change a malodorous, highly turbid, grey-to-yellow influent to an odourless, clear, light yellow effluent. A filter capacity of about 0·05 m^3/hd is adequate and a satisfactory specification for the aggregate is:

top 100 mm :   3– 6 mm
bottom 400 mm : 12–18 mm

124

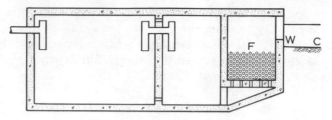

Figure 12.3 Septic tank with upflow filter. F, filter; W, effluent weir; C, effluent channel

Such a filter may be expected to operate satisfactorily without maintenance for 18–24 months when it becomes necessary to drain the filter and wash it with one or two flushes of clean water. The filter can be incorporated into the second compartment of the septic tank (Figure 12.3). Both the tank and the filter may then be cleaned at the same time. The filter effluent may be discharged into a stream or disposed of in drainfield trenches or evapotranspiration beds. Alternatively it may be used for small flood irrigation schemes (Section 13.3).

12.4 AQUA-PRIVIES

Aqua-privies are essentially septic tanks which are situated immediately below the lavatory pan or squatting plate (Figure 12·4). The liquid level in the tank

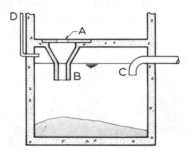

Figure 12.4 Simple aqua-privy. A, squatting plate; B, drop pipe; C, effluent pipe; D, vent pipe

must be maintained so that a water seal is formed at the vertical drop hole. Since sullage is not usually added to the tank, the water level is in practice often too low to make the desired water seal and there are the consequent problems of odour release and of fly and mosquito breeding. A solution is to divert sullage into the tank; even then additional water may be required. As a result these units are not very popular as permanent installations on individual plots. They are useful however as temporary communal facilities in military and refugee camps where there would be direct control over the maintenance of the water seal.

12.5 REFERENCES

1. Vincent, J. L., in *Proceedings of a Symposium on Hygiene and Sanitation in Relation to Housing*, Commission for Technical Cooperation in Africa, London, 1963.
*2. Malan, W. M., *CSIR Research Report No. 219*, Council for Scientific and Industrial Research, Pretoria, 1964.
*3. Cotteral, J. A. and Norris, D. M., *Journal of the Sanitary Engineering, American Society of Civil Engineers*, **95**, 715 (1969).
4. Raman, V. and Chakladar, N. *Journal of the Water Pollution Control Federation*, **44**, 1552 (1972).
*Recommended reading

Further reading

R. Laak *et al.*, 'Rational Basis for Septic Tank System Design', *Ground Water*, **12**, 348–352 (1974).

12.6 DESIGN EXAMPLE

Design a septic tank for a family of twelve which has a water consumption of 100 l/hd/d. Effluent disposal is to be by subsurface irrigation in a drainfield.
Solution
Assume that the waste flow is 80 per cent of the water consumption and that all of this enters the septic tank. Allowing 3 d retention at start-up, the tank volume is:

$$(0.08 \text{ m}^3/\text{hd d}) \times (12 \text{ people}) \times (3 \text{ d retention}) = 2.9 \text{ m}^3$$

say 2·9 m long × 1 m wide × 1 m deep with the inter-compartmental wall 1·9 m from the inlet end.

Assuming that the rate of sludge accumulation is $0.04 \text{ m}^3/\text{hd yr}$, the desludging interval is:

$$(0.08 \text{ m}^3/\text{hd d})/(0.04 \text{ m}^3/\text{hd yr}) = 2 \text{ years}$$

Alternative design procedure: Assume that the sludge accumulation rate is $0.04 \text{ m}^3/\text{hd yr}$ as above but chose the desludging interval to be *n* years (*n* may be 2, 3, 5 or 10 years). Then the tank volume is given by:

$$3 \times (0.04 \text{ m}^3/\text{hd yr}) \times (n \text{ yr}) \times (12 \text{ people}).$$

The factor 3 is introduced as the tank is one-*third* full of sludge just prior to desludging. Thus if *n* were chosen as 3 years, the tank volume would be 4·3 m³. The retention time at start-up would be:

$$\frac{(4.3 \text{ m}^3)}{(0.08 \text{ m}^3/\text{hd d}) \times (12 \text{ people})}$$

i.e. about 4·5 d, which is satisfactory.

Drainage trench design: Assume that the infiltration rate is 10 l/m² d and that the trenches have an effective depth of 0·7 m. The sidewall area required is given by the ratio:

$$\frac{\text{effluent flow (l/d)}}{\text{infiltration rate (l/m}^2\text{ d)}}$$

$$= \frac{80 \times 12}{10} = 96 \text{ m}^2$$

Remembering that the trench has two sides, the total trench length is given by:

$$\frac{(\text{infiltration area})/2}{\text{effective depth}}$$

$$= \frac{96}{2 \times 0.7} = 70 \text{ m}$$

Note that in calculating the infiltration area only the sidewalls are considered. This is because the bottom of the trench becomes rapidly clogged, leaving the sidewalls as the only effective infiltration surfaces.

13

Effluent Re-use

13.1 RE-USE AND RECLAMATION

The general scarcity of water in the tropics and subtropics and the high costs of developing new water supplies are the two major factors responsible for the increasing recognition of the need to conserve water resources by effluent *re-use* (e.g. for aquaculture or irrigation) or by effluent *reclamation* to produce a water suitable for industry (e.g. cooling water) or even one of drinking water quality. The re-use of both raw and treated wastewaters for irrigation has been widely practised for many years; more recently attention has been given to aquaculture and the municipal and industrial re-use of effluent.

13.2 AQUACULTURE

Aquaculture means 'water farming', just as agriculture means 'field farming'; it is the growing of plants and animals in water for their eventual harvesting as food, either for man or domestic animals; rice growing is the classic example of aquaculture. However, as sanitary engineers we are concerned not with rice, but rather with algae, fish and ducks. The dense algal blooms in waste stabilization ponds not only provide oxygen for the bacterial oxidation of the influent sewage, they are also a valuable food source, being approximately 50 per cent protein. The growth of algae in ponds is a highly efficient process with protein yields far in excess of those commonly found in conventional agriculture (Table 13.1). The algae may be harvested from maturation pond effluent by one of several tertiary treatment processes (Section 11.2) and then used as an animal food supplement. Sewage-grown algae have been successfully fed to chickens, pig, cattle and sheep.[1] Often however there is neither the money nor the skill to install operate and maintain tertiary treatment processes. In such cases the algal protein in waste stabilization ponds is most conveniently exploited by growing algae-eating fish in the maturation ponds.[2,3] The tilapia *Sarotherodon mossambica* is particularly tolerant of high algal densities and grows extremely well in maturation ponds; moreover it is very good to eat. Other fish which have been grown in maturation ponds include carp (*Catla catla, Laboe rohita* (Frontispiece)) channel catfish (*Ictalurus punctatus*) and mosquito fish (*Gambusia* spp.). Fish yields, particularly of tilapia, may be

128

Table 13.1 Protein yields of various agricultural and aquaculral crops*

Crop	Protein yield (kg/ha year)
Soya bean	650
Maize	270
Wheat	150
Rice	55
Algae	82 000[†]

*Adapted from McGarry.[1]
†Yield from high-rate ponds.

increased by the introduction of sterile hybrids,[4] but this requires the expert
attention of an experienced fish-farmer.

In Papua New Guinea a very simple but extremely effective combination
of aquaculture and agriculture has been developed:[5] maturation ponds are
used for raising both fish and ducks and the final effluent is used for the irrigation
of high quality vegetables which are grown in gravel rather than soil, a horti-
cultural practice known as 'hydroponics' (Figure 13.1).

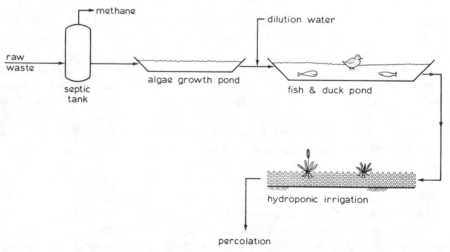

Figure 13.1 Integrated waste disposal system as developed in Papua New Guinea.
Methane is collected from the septic tank and used as a domestic fuel. The algae growth
pond is a facultative stabilization pond and designed accordingly. Fish and ducks are
reared in a maturation pond to which groundwater or river water is added to increase
the pond volume and replace evaporation losses. The pond effluent is used to irrigate a
vegetable garden; any remaining liquid is run to waste or allowed to percolate through
the soil

Clonorchiasis

This serious disease of the liver is caused by the parasitic trematode worm *Clonorchis sinensis* which has two intermediate hosts, snails (*Bithynia* spp.) and fish (carp). It is confined to the Far East, especially southern China, where its high indidence is due to the extensive practice of fertilizing fish ponds with sewage and the local preference for eating raw fish.

Regular pond maintenance is essential to prevent the development of a stable snail population. If the banks are clear of vegetation, then the snail is deprived of its habitat and therefore the parasites in the raw sewage die owing to the absence of their first host.

13.3 AGRICULTURE RE-USE

The re-use of effluent in agricultural practice is mostly restricted to crop irrigation; stock watering is not recommended because of the health risk to the animals. The two major considerations which govern the suitability of an effluent for irrigation are its chemical quality and the potential risks to public health arising from its use.

Health aspects

Careful consideration must be given to the hazards to health that may arise from the re-use of wastewater; in some circumstances they may preclude such re-use. The hazards to health can be reduced by the treatment of wastewater before use in agriculture, but such treatment is often impossible for economic reasons. The health of the public can often best be protected by restricting irrigation with wastewater to certain crops, so that municipal wastewater will not come into contact with any part of a plant used as food for man, particularly if it is eaten uncooked. Broad irrigation of land bearing fruit trees is usually safe, whereas spray irrigation is not. It is clearly desirable to restrict the irrigation with wastewater of crops used for fodder and industrial purposes; if fodder crops or parture are so irrigated, there is a risk that cattle will be infected with larvae of the beef tapeworm, *Taenia saginata*.[6]

Irrigation

Irrigation areas should be farmed in one unit to which only trained workers have access. Under no circumstances should such areas be in the hands of smallholders, where a dual system of water-supply pipes would be necessary. Sooner or later cross connexions would occur, or children would drink from the wrong tap.[7]

It is now generally accepted that it is extremely poor practice to use raw sewage for irrigation. This consideration is based solely on the health risks to those working on such an irrigation scheme and to the general public which may consume contaminated produce. Surveys in India have shown that 80 per cent of workers on 'raw sewage farms' were infected with intestinal parasites as compared with 28 per cent in the control populations; these workers were also found to be significantly more susceptible to anaemia, skin disorders and diseases of the respiratory and intestinal tracts.[8] The health of farm workers receives substantial protection if effluent is used for irrigation rather than the raw waste. Health risks to the general public are best minimized by restricting the use of even effluent to the irrigation of industrial or fodder crops. Although effluent may be used to irrigate certain crops intended for human consumption, it should *never* be used to irrigate those that are eaten raw. Rigorous control by a Government agency, backed by the necessary legal authority, is required; detailed regulations have in fact been drawn up in several countries.[9]

Bacteriological quality

If the effluent is to be used for 'restricted' irrigation (i.e. for industrial or fodder crops) the FC density should be $< 5000/100$ ml (which is the general minimum standard—Section 3.4). American[10] and South African[11] practice requires $FC < 1000/100$ ml. However, if the effluent is to be used for unrestricted irrigation, a much higher standard is necessary: $FC < 100/100$ ml.[12]

Chemical quality

Irrigation water must not contain any compound toxic to the crop under cultivation. The chemical suitability of an effluent for irrigation is judged mainly by three parameters—its electrical conductivity, sodium absorption ratio and boron content.

Electrical conductivity is a convenient measure of dissolved salts and hence of the 'salinity hazard' to the crop. Excessive salinity reduces a plant's osmotic activity and so prevents its absorption of both water and nutrients from the soil.

Sodium absorption ratio is a measure of the alkali or 'sodium hazard' to the crop. Sodium ions tend to become adsorbed on to the soil particles, displacing magnesium and calcium ions as they do so; when the sodium concentration is high the exchange of Na^+ for Ca^{2+} and Mg^{2+} results in a soil with poor internal drainage which restricts the circulation of air and water when wet and which usually forms hard and unmanageable clods when dry. SAR is defined as:

$$\frac{[Na]}{\frac{1}{2}([Ca] + [Mg])^{1/2}}$$

where [Na], [Ca] and [Mg] are the concentrations of sodium, calcium and magnesium in milli-equivalents per litre (the concentration in meq/l is obtained by multiplying the concentration in mg/l by 0·044, 0·050 and 0·082 for Na, Ca and Mg respectively).

The permissible values of conductivity and SAR are interdependent (Figure 13.2); generally effluents of conductivity < 100 mS/m (at 25 °C) and SAR < 15 are satisfactory for most crops.

Boron at a concentration $> 0·5$ mg/l is toxic to citrus and deciduous fruits and nuts, although concentrations < 2 mg/l are generally satisfactory for most crops.

131

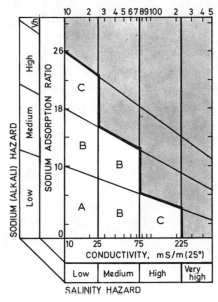

Figure 13.2 Classification of irrigation
waters based on conductivity and SAR.
Waters in regions A and B are acceptable
for almost all irrigation purposes; those
in regions C should be avoided wherever
possible and those in the shaded area
should not be used at all [Adapted from
*United States Department of Agriculture
Handbook No. 60*, 1954]

Table 13.2 Effluent quality of three facultative waste stabilization
ponds in Israel*

		Pond A	Pond B	Pond C
Chloride	(Cl)	532[+]	104	408
Bicarbonate	(HCO_3^-)	176	320	677
Sulphate	(SO_4^-)	44	20	15
Boron	(B)	nil	0·1	0·3
Phosphorus	(P)	1·6	nil	9
Sodium	(Na)	194	62	215
Calcium	(Ca)	59	32	73
Magnesium	(Mg)	292	47	122
SAR		3·4	1·6	3·9
pH		7·0	7·1	6·9

*From Watson.[7]
[+]Concentrations in mg/l (except SAR and pH).

132

The effluent quality of some facultative waste stabilization ponds in Israel is given in Table 13.2.

Methods of irrigation

There are basically three methods of distributing irrigation waters—subsurface, surface and spray irrigation. Subsurface irrigation is generally only used for the disposal of septic tank effluents (Chapter 12). Surface irrigation is usually in furrows and is used for crops grown in regular rows (e.g. maize). As a guide 1 ha of irrigation area is required for a flow of 25 m^3/d.[7]

Flood irrigation is an alternative method of surface application: the irrigation area (which should be enclosed by small embankments) is flooded at regular intervals to a depth of 100–300 mm and the effluent allowed to percolate down slowly through the soil. As a general rule-of-thumb the volume of water required for flood irrigation schemes is approximately 10 mm/ha d: thus if the waste flow is 80 l/hd d an irrigation area of about 8 m^2 is required for each person contributing to the waste flow. This method of irrigation is particularly suited to small schemes for the disposal of septic tank effluent in areas covered with an impermeable soil such as black cotton soil. It is advantageous to cultivate essential oil bearing plants such as citronella (*Cymbopogon winterianus*) and peppermint (*Mentha arvensis*) because they are labour intensive cash crops which grow well on black cotton soil.[13,14]

Spray irrigation is a common alternative to surface irrigation and has the advantages of greater versatility and control. It requires less maintenance but is more expensive. In a properly operated scheme water wastage is minimal, which is a further advantage especially in arid regions. There are two ways in which the irrigation water is commonly disturbed: double-nozzle revolving sprinkler heads spaced along a portable pipe and rotating spray rigs (Figures 13.3 and 13.4).

Figure 13.3 Sprinkler head attached to portable aluminium pipe [Courtesy of McDowell Mfg Co.]

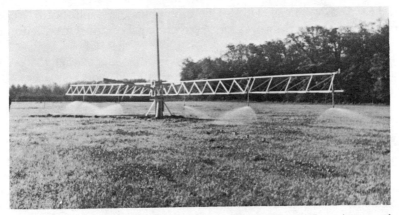

Figure 13.4 Rotating spray rig, 'Aquamast'. Note that the effluent is sprayed *downwards* to minimize the possibility of wind-borne contamination. This model serves an area of 0·1 ha; other models are available for areas up to 16 ha
[Courtesy of McDowell Mfg Co.]

The Muskegon scheme[15]

The integration of wastewater treatment and agricultural management is exemplified by the $42 million spray irrigation project adopted by the county

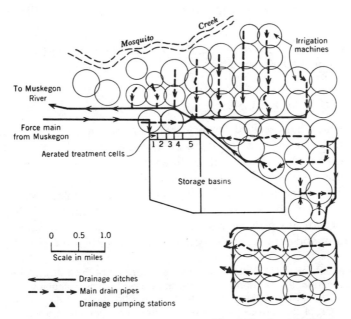

Figure 13.5 The Muskegon scheme. The irrigation machines are similar to that shown in Figure 13.4 [From Chaiken *et al.*[15]]

of Muskegon in Michigan, USA. This ambitious and sophisticated scheme, the largest of its kind in the world, has seven principal components (Figure 13.5):

(1) The collection and transport of 164 000 m^3/d of raw sewage (55 per cent industrial) which is pumped 18 km to the treatment works adjacent to the irrigation area.
(2) Three aerated lagoons (each 3·24 ha × 4·6 m) which reduce the BOD by 70–90 per cent; total retention time is 3 d.
(3) Two large storage lagoons (2·7 m deep) which are used to settle the bioflocs and to store the effluent when it cannot be used for irrigation (in northern USA a retention time of 4 months is required during the winter when the ground is frozen and thus unable to accept the effluent; such a large storage area would not be required in hot climates).
(4) Chlorination of the effluent from the outlet lagoon to produce < 1000 FC/100 ml.
(5) Distribution of the chlorinated effluent to 55 rotating spray rigs for the irrigation of 4000 ha of farmland.
(6) Further treatment of the effluent as it percolates through the soil.
(7) Collection of the now fully treated effluent in a system of underdrains and final disposal to the headwaters of local streams.
The anticipated treatment efficiency is remarkably high:

Removal of BOD$_5$	99%
Suspended solids	99%
Phosphorus	90%
Nitrogen	76%

Virus removal is expected to be 100 per cent and the destruction of faecal bacteria to be > 99·99 per cent.

13.4 MUNICIPAL RE-USE

Effluent which has received tertiary treatment which includes chlorination is suitable for watering municipal parks and golf courses and street flushing; it is often cheaper to use effluent for these purposes than fully treated drinking water. Effluent can also be used in municipal fish ponds to replace evaporation losses and provide nutrients. Effluent reclamation to produce a water of potable quality has received extensive investigation in Namibia and South Africa. It is now technically possible and economically feasible to reclaim maturation pond effluent and even sewage which has received only primary settling:

The reclamation plant [at Windhoek, Namibia] consists of the following unit processes: foam fractionation, flocculation/flotation using excess lime as coagulant, ammonia stripping, stabilization, carbon absorption, filtration, and breakpoint chlorination.
    The function of the various process units cannot be regarded as discrete with respect to the removal of the various constituents present. Synthetic detergent is removed in the flotation, foam fractionation and absorption stages. Ammonia is greatly reduced in the

stripping process, but residuals can only be removed completely by applying breakpoint chlorination. From a bacteriological and virological point of view, both excess lime and break-point chlorination play a major role in securing sterilized water. Integration of the units also provides a multiple barrier system consisting of chemical oxidation, absorption and physical separation of the viruses and bacteria. The optimum functioning of the reclamation plant is, therefore, dependent on the manner in which the individual process units are combined.[16]

It is preferable to reclaim settled sewage because the nitrate content of the reclaimed water in negligible; in contrast water reclaimed from maturation ponds has a nitrate concentration of 50–100 mg/l (as NO_3^-) which is hazardous to the health of newborn infants.[17,18]

13.5 INDUSTRIAL RE-USE

The re-use of water within industry is often practised as a means of minimizing water charges, the most common example being the re-use of the wastewater of one process as cooling water for another. Sewage effluent has been used as industrial cooling water but this practice is not widespread as a result of the troublesome growth of slime organisms in pipes and cooling towers.

Water shortages are particularly serious in industry. In Bombay one large industrial company, unable to obtain sufficient water from the municipal supply, treats 5000 m³/d of sewage effluent by flocculation, sedimentation, rapid sand filtration, softening and chlorination to produce a water suitable for general industrial use.[19] In the same city, process water for the air-conditioning plant in several multistoried office blocks is obtained from the 100–250 m³/d of sewage produced in each building. The treatment plant, situated in the basement of the building, consists of a conventional activated sludge process followed by flocculation with alum, pressure filtration and softening.[19]

13.6 REFERENCES

1. McGarry, M. G., in *Proceedings of a Seminar on Water Supply and Wastewater Disposal in Developing Countries*, Asian Institute of Technology, Bangkok, 1971.
2. Hey, D., *Proceedings of the International Association Theoretical & Applied Limnology*, 12, 737 (1953).
3. Knapp, C. E., *Environmental Science & Technology*, 5, 122 (1971).
4. Hickling, C. F., *FAO Fisheries Report No. 44*, 4, 1 (1968).
5. Chan, G., in *Workshop on Waste Recycling Systems*, Department of Agriculture, Stock and Fisheries, Papua New Guinea, 1973.
*6. *Water Pollution Control in Developing Countries (Technical Report Series No. 404)*, World Health Organization, Geneva, 1968.
*7. Watson, J. L. A., *Proceedings of the Institution of Civil Engineers*, 22, 21 (1962).
8. *Health Status of Sewage Farm Workers (Technical Digest No. 17)*, Central Public Health Engineering Research Institute, Nagpur, India, 1971.
*9. Shuval, H. I., *Water Research*, 1, 297 (1967).
10. Geldreich, E. E. and Bordner, R. H., *Journal of Milk and Food Technology*, 34, 184 (1971).
11. Meiring, P. G. J. *et al.*, *CSIR Special Report WAT 34*, Council for Scientific and Industrial Research, Pretoria, 1968.

*12. *Reuse of Effluents: Methods of Wastewater Treatment and Health Safeguards (Technical Report Series No. 517)*, World Health Organization, Geneva, 1973.

13. *Sewage Farming (Technical Digest No. 3)*, Central Public Health Engineering Research Institute, Nagpur, 1970.

14. Campbell, P. A. and Mara, D. D., in *Proceedings of a Seminar on Sewage Treatment*, University of Nairobi, Kenya, 1973.

15. Chaiken, E. I. *et al.*, *Civil Engineering ASCE*, (5), 49 (1973).

*16. Stander, G. J. *et al.*, *Water Pollution Control*, **70**, 213 (1971).

17. Stander, G. J. and van Vuuren, L. R. J., *Journal of the Water Pollution Control Federation*, **41**, 355 (1967).

18. *International Standards for Drinking Water*, 3rd edition, World Health Organization, Geneva, 1971.

19. Chhabria, N. D., *Chemical Processing and Engineering*, (5), 62 (1971).

*Recommended reading

Further reading

Culp, R. L. and Culp, G. L. *Advanced Wastewater Treatment*, Van Nostrand Reinhold, New York, 1971.

Gotaas, H. B., *Composting: Sanitary Disposal and Reclamation of Organic Wastes*, World Health Organization, Geneva, 1956.

Payne, I., 'Tilapia—a fish of culture', *New Scientist*, **67**, 256 (1975).

Radebaugh, G. H. and Agersborg, H. P., 'The economic value of treated sewage effluent in wildlife conservation, with special reference to fish and waterfowl', *Transactions of the American Fisheries Society*, **64**, 443 (1934).

Schroeder, G. L., 'Some effects of stocking fish in waste treatment ponds', *Water Research*, **9**, 591 (1975).

Workshop on Waste Recycling Systems, Department of Agriculture, Stock and Fisheries, Papua New Guinea, 1973.

14

Collection and Treatment of Nightsoil

14.1 ECONOMIC CONSIDERATIONS

Waterborne sewerage systems are highly capital intensive. In developing countries they are often too expensive for the community which they are intended to serve; for example in Belize it was estimated that if full facilities for waterborne sewerage were installed the revenue collected (based on estimates of the householders' ability to pay) would, after deduction of operation and maintenance costs, be sufficient to amortize only one-eighth of the capital costs involved.[1] In tropical developing countries where capital is scarce but labour plentiful and relatively cheap, labour intensive schemes are economically and socially more advantageous. In urban areas (where pit latrines and other essentially rural methods of excreta disposal are inappropriate) the manual collection of nightsoil (human faeces and urine) has much lower capital requirements and much higher labour and maintenance costs than waterborne sewerage. With careful planning, nightsoil collection can be as hygienic as waterborne sewerage. It is therefore a feasible solution, both technically and financially, to the enormous problem of urban excreta disposal.

14.2 TRADITIONAL METHODS

Nightsoil is collected from either buckets or nightsoil vaults situated immediately below the toilet. This process is rarely done with any semblance of hygiene; for example, in one African country:

The collection and disposal of nightsoil from bucket lavatories is usually nauseating. Although in some cases the buckets are manually carried long distances to the disposal ground, the usual practice is to empty the buckets into handcarts, each comprising an empty drum supported horizontally across two wheels; when full, the handcarts are dragged away and either buried or emptied into a sewer, septic tank or local depression. Only rarely are the buckets and handcarts washed after use; spillage of nightsoil is frequent and health hazards are alarmingly obvious. The bucket lavatories are rarely disinfected. They are almost always unhygienic, offensive and usually surrounded by insects, many of which help spread human diseases; sometimes a degree of cleanliness is unintentionally achieved by keeping poultry which devour these insects.[2]

In India the situation is similar:

138

It is common to see a scavenger moving with a heavy load of nightsoil on his/her head in a bamboo basket or a leaky drum, the contents trickling over the carrier.[3]

The health hazards resulting from such dreadful nightsoil collection practices are patently infinate. In order to reduce them by several orders of magnitude, a special nightsoil wheelbarrow has been developed and subjected to field trials in central India.[3]

In Asia untreated nightsoil is the traditional fertilizer for vegetable crops and fish ponds. It is usually collected from nightsoil vaults manually by dipper and bucket (Figure 14.1). This is an unhygienic process as it is impossible to avoid spillages and vermin usually have ready access to the vaults. Yet agricultural demand for nightsoil is high; this has both advantages and disadvantages—for example, in Tainan:

The municipality regards nightsoil as a monetary asset and is often annoyed by thieves who steal excreta, thereby reducing the municipality's income which it uses to offset nightsoil collection costs. In essence there exists a black market in nightsoil.[4]

Figure 14.1 Nightsoil collection by dipper and bucket in Taiwan [Courtesy of Dr M. G. McGarry]

14.3 MODERN METHODS

Modern methods must avoid the extreme health hazards associated with the traditional methods of nightsoil collection and disposal. Yet at the same time they must be inexpensive and simple.

Collection

Watertight, vented nightsoil vaults or tanks used in conjunction with a water-seal toilet of low water consumption (about 1 l/flush) are cheap and easy to install. They are emptied under vacuum once every 2–3 weeks into either large conservancy tankers or smaller electric or ox-pulled nightsoil carts (these carts are essentially small conservancy tankers; they would incorporate manually operated vacuum pumps). This system,[4] which is very similar to that used extensively in Japan,[5] is free from odour and avoids the hazards from spillage and insects. Communal toilet blocks ('comfort stations') should be provided in areas of high population density where the residents cannot afford to build nightsoil toilets in their own homes. These *must* be kept scrupulously clean in order to ensure their continued use.

Treatment and re-use

After collection the nightsoil is transported to a central depot where it is treated so that subsequent health hazards are eliminated. Pasteurization by steam injection is a suitable method of treatment.[4] The nightsoil is heated to 80 °C for 30 minutes. This ensures the destruction of the numerous entero-pathogenic bacteria and parasite eggs present. After this treatment the nightsoil is harmless and may be safely applied to vegetable crops and fish ponds.

An alternative method of nightsoil treatment is anaerobic digestion.[6] This has the advantage of a net energy output in the form of methane gas. The digested sludge is used as a fertilizer and, to prevent water loss, it is most conveniently applied to the land in liquid form. In Kerala State, India, there are two full-scale municipal nightsoil digesters serving populations of 15 000 and 20 000. The digester capacity is $0 \cdot 03 - 0 \cdot 06 \, m^3/hd$. Animal wastes (especially cow and pig dung) are also very amenable to anaerobic digestion and yield large quantities of gas.

Nightsoil can also be treated in facultative waste stabilization ponds.[7] Water lost by evaporation must be regularly replaced and the pond must be properly maintained. A special concrete inlet ramp is provided which permits the nightsoil to be sluiced into the pond by a jet of water. The area required for the nightsoil pond is calculated from equation 7.16 on the assumption that the nightsoil BOD_5 contribution is 22 g/hd d (values for faeces + urine from Table 1.1). The algae produced in the nightsoil pond may be harvested for use as livestock feed or the pond effluent led to a fish pond (Figure 14.2).

In Lagos[8] the nightsoil from 14 000 houses is collected in pails and taken to one of several collection depots from where it is taken to the treatment works by tanker. The volume of nightsoil collected is 180 m^3/d and the BOD_5 load 8300 kg/d. At the treatment works (Figure 14.3) the nightsoil is screened,

140

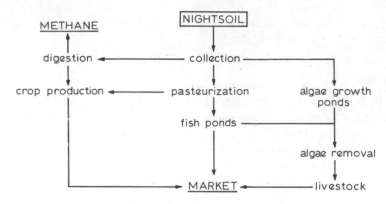

Figure 14.2 Nightsoil treatment alternatives

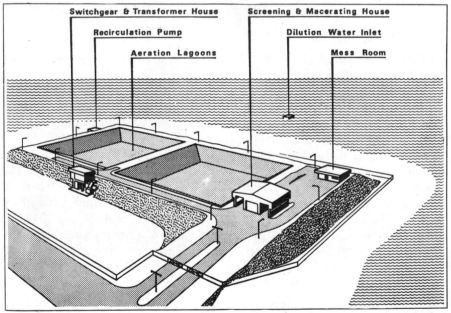

Figure 14.3 Aerated lagoons for nightsoil treatment in Lagos [From Hindhaugh[8]]

diluted with make-up water and final effluent, macerated and finally treated in two aerated lagoons (each 55 m square × 3 m deep and having four 75 h.p. aerators). The treated waste is discharged into Lagos lagoon some 120 m below low water. An interesting design requirement for this scheme (which is the largest of its type in the world) was that the works should be capable of accepting waterborne sewage at some future date with the minimum of modification and expense.

14.4 REFERENCES

1. McGarry, M. G., personal communication, 1974.
2. Holland, R. J., personal communication, 1973.
3. *Nightsoil Wheelbarrows (Technical Digest No. 32)*, Central Public Health Engineering Research Institute, Nagpur, 1972.
4. McGarry, M. G., in *Proceedings of a Symposium on the Role of the Engineer in Environmental Pollution Control, Kuala Lumpur*, Institute of Engineers, Malaysia, 1972.
5. Pradt, L. A., *Water Research*, **5**, 507 (1971).
6. *Digestion of Nightsoil and Cow Dung (Technical Digest No. 8)*, Central Public Health Engineering Research Institute, Nagpur, 1970.
7. Shaw, V. A., *Public Health, Johannesburg*, **63**, 17 (1963).
8. Hindhaugh, G. M. A., *The Consulting Engineer*, (9), 47 (1973).

14.5 DESIGN EXAMPLE

Design a facultative stabilization pond to treat the nightsoil from a population of 10 000. The design temperature is 18 °C.

Solution

Assume that the BOD_5 contribution = 22 g/hd d.

BOD_5 load = $22 \times 10^{-3} \times 10\,000 = 220$ kg/d.

Design loading on pond is given by equation 7.16:

$$\lambda_s = 20\,T - 120 = (20 \times 18) - 120$$

$$= 240 \text{ kg/ha d}$$

Therefore pond area = 220/240 = 0·92 ha.

Choose the depth as 1·2 m (Section 7.8).

Appendix 1

Measurement of BOD, COD and FC

Chemical oxygen demand[1,2]

Principle of method

Organic compounds are oxidized to carbon dioxide and water by a boiling acid dichromate solution:

$$C_aH_bO_c + dCr_2O_7^{2-} + 8\,dH^+ \xrightarrow{\text{heat}} aCO_2 + eH_2O + 2\,cCr^{3+}$$

where $d = \frac{2}{3}a + \frac{1}{6}b - \frac{1}{3}c$
$e = \frac{1}{2}b + 4\,d.$

The amount of dichromate which has participated in this reaction ($\equiv$ COD of sample) is then determined in a colorimeter at a wavelength of 600 nm.

Reagents required

(1) N/4 potassium dichromate: 12·259 g/l analytical grade $K_2Cr_2O_7$ (previously dried at 103 °C for 2 h); to eliminate interference caused by up to 5 mg/l nitrite-nitrogen incorporate 100 mg/l sulphamic acid (or *pro rata*).
(2) Analytical grade sulphuric acid (relative density = 1·84).
(3) Mercuric sulphate: analytical grade crystals (required to eliminate inter-ference by up to 1000 mg/l chloride).

Apparatus

(1) Conical (Erlyenmeyer) flasks, capacity 250 ml, fitted with 24/29 ground glass socket (e.g. Jobling-Quickfit[3] FE 250/3).
(2) Liebig condensers, effective length 300 mm, fitted with 24/29 ground glass cone (e.g. Jobling-Quickfit[3] CX 1/23).
(3) Electric hotplate or bunsen burner and stand.
(4) Colorimeter (e.g. Walden Precision Apparatus[4] model CO-35).

Procedure

(1) Add to one conical flask 20 ml of the sample and to another 20 ml of distilled water. Add to each:

> 0.2 g solid mercuric sulphate
> 10·0 ml standard potassium dichromate solution
> a few undrilled glass balls (4 mm dia.) or porcelain boiling chips

(2) Fit a Liebig condenser to each flask and *slowly* pour in from the top of the condenser 30 ml of the sulphuric acid–silver sulphate solution. *Now thoroughly mix the contents of the flask.*

(3) Place flask and condenser on hotplate, connect water supply to condenser, and leave to boil for 2 h. (The use of a condenser permits the flask contents to boil without loss of volume and without loss of any volatile organic compounds present.)

(4) Remove flasks from heat, allow to cool, and pour their contents into two optically matched 125×16 mm Pyrex test tubes. Place the tube containing the distilled water blank in the colorimeter and adjust the meter to read zero optical density at a wavelength of 600 nm. Place the tube containing the sample in the colorimeter and read the optical density.

(5) Determine the COD from the optical density by reference to a calibration graph which has been prepared with standard solutions of potassium hydrogen phthalate[5] (use solutions of 212·5, 425·0, 637·5 and 850·0 mg/l for COD's of 250, 500, 750 and 1000 mg/l respectively). These standard solutions are analysed as described in steps (1)–(4) above.

Sample preservation

If concentrated sulphuric acid is added to the sample at the rate of 2 ml/l, the sample may be kept at room temperature for up to 10 d.

Biochemical oxygen demand

There are three principal methods for measuring BOD—the standard bottle test, the ΔCOD procedure and the Hach manometric BOD meter.

Hach BOD meter[6]

This meter (Figure A1.1) works on the principle that the air pressure in a closed bottle partially filled with an aerobic bacterial culture (such as a BOD sample) decreases as the bacteria grow (i.e. as the waste is oxidized). The fall in air pressure is proportional to the BOD of the sample and is monitored by an expanded limb U-tube manometer which is calibrated in mg/l BOD_5. To avoid errors introduced by carbon dioxide evolution a CO_2 absorbent (40 per cent potassium hydroxide) is provided. The meter, which is designed for five samples, is essentially a row of five magnetic stirrers with a U-tube manometer positioned

Figure A1.1 Hach BOD meter [Courtesy of Hach Chemical Co.]

by each stirrer. It is extremely simple to operate and it has the advantage that the BOD curve (Figure 4.1) is obtained merely by taking a reading each day.

Standard bottle test

The principle of this method is to measure the change over 5 d in DO concentration in a stoppered bottle completely filled with the waste or a dilution of it. Dilution is necessary if the BOD is more than the solubility of oxygen in the waste (approximately 8–9 mg/l at 20 °C). A special dilution water is used which is buffered to pH 7·2 and contains essential inorganic nutrients. A suitable procedure is:

(1) Make up BOD dilution water by adding to each litre of air-saturated distilled water:

    2 ml ferric chloride solution ($0·25$ g/l $FeCl_3.6\,H_2O$)
    2 ml calcium chloride solution ($36·4$ g/l $CaCl_2.2\,H_2O$)
    2 ml magnesium sulphate solution ($22·5$ g/l $MgSO_4.7\,H_2O$)
    2 ml phosphate buffer ($8·5$ g KH_2PO_4, $21·75$ g K_2HPO_4, $33·4$ g $Na_2PO_4.7\,H_2O$ and $1·7$ g NH_4Cl in 1000 ml distilled water)

Distilled water may be saturated with air by bubbling air through it or by stirring it on a magnetic stirrer for 1–2 h. If the sample contains $< 10^3$ bacteria/ml, 2 ml of settled sewage should be added to each litre of dilution water. This 'seeding' procedure provides a sufficiently large bacterial population to enable bio-oxidation to occur without delay; it is usually only required when industrial effluents (which are often deficient in bacteria) are being examined.

Table A1.1 BOD sample dilution ranges

Volume of sample* (ml)	Dilution (%)	Range of BOD$_5$[+] (mg/l)
250	100	0– 8
100	40	5– 20
50	20	10– 40
25	10	20– 80
10·0	4	50– 200
5·0	2	100– 400
2·0	0·8	250–1000
1·0	0·4	500–2000
0·50	0·2	1000–4000

*to be pipetted into 250 ml BOD bottle.
[+]based on a minimum O_2 depletion in the BOD bottle of 2 mg/l and a minimum residual DO concentration of 1 mg/l (i.e. a maximum O_2 depletion of 8 mg/l).

(2) Bring the sample to 20 °C and aerate until saturated (about $\frac{1}{2}$ h).
(3) Estimate the BOD of the sample and select three dilutions from Table A1.1. For each dilution pipette the appropriate sample volume into each of three 250 ml glass-stoppered reagent vessels (e.g. Jobling-Pyrex[3] 1520/06). Fill each bottle with dilution water to within 1 cm of the top of the neck and re-stopper; there should be no bubbles of air trapped in the bottle. (The dilution water should be *siphoned* into the BOD bottle via a length of flexible rubber tubing the end of which is kept just *below* the surface of the liquid in the bottle; in this way the bottles can be filled in a uniform, reproducible manner which avoids aeration.)
(4) Three bottles are similarly filled with dilution water only, to serve as 'blanks'.
(5) All twelve bottles are incubated at 20 °C for 5 d. After this period the dissolved concentration in each is measured (see below).
   The BOD$_5$ is calculated as:

$$(DO_b - DO_d) \times d$$

where DO_b = mean DO concentration in the blanks, mg/l
   DO_d = mean DO concentration in the bottles containing the sample at dilution d, mg/l
   d = dilution factor, = vol. of bottle/ml sample (choose the dilution factor for which the DO concentration lies in the range 1–7 mg/l; if two of the three dilution factors satisfy this criterion choose the one with the *lower* DO concentration).
(6) If the dilution factor is > 10 per cent, five bottles should set up, rather than three. Two of these are used to determine the initial DO concentration and the BOD$_5$ is calculated as:

$$(DO_{d0} - DO_{d5}) \times d$$

where DO_{d0} = mean DO concentration at zero time
  DO_{d5} = mean DO concentration after incubation.
For dilution factors < 10 per cent this procedure is not considered necessary since it is assumed that the initial DO concentration in the BOD bottle is the same as that in the blank.

Measurement of dissolved oxygen[7]

This method is similar to that given in *Standard Methods*[1] but is more accurate as it prevents loss of part of the iodine (which is produced in proportion to the DO originally present). Interference due to 20 mg/l nitrite-nitrogen and 3% ferric iron is eliminated by the sodium azide and phosphoric acid respectively.
  The following reagents are required:

(1) Manganous sulphate solution: dissolve 480 g $MnSO_4.4H_2O$ *or* 400 g $MnSO_4.2H_2O$ *or* 364 g $MnSO_4.H_2O$ in distilled water and dilute to 1000 ml.
(2) Sodium azide solution: dissolve 37·5 g sodium azide in 1000 ml distilled water.
(3) Alkali–iodide solution: dissolve 400 g sodium hydroxide in 560 ml distilled water and, while still hot, add 900 g sodium iodide. When cool, dilute to 1000 ml, store overnight and decant into 100–250 ml bottles.
(4) 90% orthophosphoric acid.
(5) N/40 sodium thiosulphate: dissolve 6·205 g $NA_2SO_3.5H_2O$ in freshly boiled and cooled distilled water. Preserve with 0·3–0·5 g/l sodium hydroxide.
(6) Starch solution: dissolve 5 g soluble starch in 1000 ml boiling distilled water, allow to boil for 2–3 minutes and dispense into 100–250 ml bottles. Preserve with a few drops of toluene.

The procedure is as follows:

(1) Remove the stopper from the BOD bottle and add, in the following sequence:

> 2 ml manganous sulphate solution
> 2 ml sodium azide solution
> 2 ml alkali–iodide solution

In each case the tip of the pipette should be 2–5 cm below the neck of the bottle so that the 2 ml quantities are discharged directly into the bulk of the contents. Replace the stopper (some overflow will occur, so do this in a basin).
(2) Immediately after the addition of the alkali–iodide reagent a brown flocculent precipitate forms. Shake the bottle to suspend this precipitate, allow it to settle half way down the bottle, shake again and allow to settle. This procedure ensures that all the DO present reacts with the reagents.

At this stage all the DO has been converted to manganese dioxide:

$$Mn^{2+} + 2OH^- + \tfrac{1}{2}O_2 \rightarrow MnO_2 + H_2O$$

(3) When the floc has settled the second time, unstopper the bottle and add 2 ml orthophosphoric acid, re-stopper and shake. The floc will dissolve and the bottle contents turn yellow.

Under the resulting conditions of low pH in the bottle, the manganese diozide oxidizes the iodide present to liberate free iodine:

$$MnO_2 + 2\,I^- + 4\,H^+ \rightarrow Mn^{2+} + 2\,H_2O + I_2$$

(4) Titrate 205 ml of the bottle contents with N/40 (0·025N) sodium thio-sulphate until pale yellow. Add 1 ml starch solution and add thiosulphate drop by drop until the colour *suddenly* changes from blue to colourless. Ignore any subsequent return of the blue colour. Calculate the DO concentration, mg/l:

$$DO = (\text{ml of thiosulphate used}) \times \frac{0·025}{N}$$

where N = normality of thiosulphate solution.

The volume taken for the titration allows for the dilution of the sample by the 6 ml of the first three reagents and is equivalent to 200 ml of the sample since 200 × 250/(250 − 6) = 205 (here 250 is the bottle volume); no allowance is made for the 2 ml of phosphoric acid since it does not displace any of the DO which is then present as MnO_2. Since 1 ml of N/40 thiosulphate is equivalent to 0·2 mg DO and since there are five 200 ml volumes in a litre, the DO concentration is (0·2 × 5) × ml of thiosulphate used. If the normality of the thiosulphate is not exactly N/40, a correction is made. The normality may be determined by titrating 25·00 ml N/40 potassium dichromate (prepared by dilution of the N/4 reagent used in the COD test) with the thiosulphate solution, adding starch when pale yellow as above. The normality is given by:

$$N = 0·025 \times \frac{25}{\text{ml thiosulphate used}}$$

Measurement of BOD as ΔCOD[8]

BOD_n is measured as the difference, ΔCOD, of the initial COD and the COD remaining after aerobic incubation at 20 °C for n d. This is possible because the COD of a waste is the oxygen demand of the biodegradable fraction of its organic compounds ($= BOD_u$) plus that of the non-biodegradable fraction ($= NBDG$). Therefore at zero time and after n d:

$$COD_0 = (BOD_u)_0 + (NBDG)_0$$
$$COD_n = (BOD_u)_n + (NBDG)_n$$

$$\Delta COD_n = (BOD_u)_0 - (BOD_u)_n + (NBDG)_0 - (NBDG)_n$$

Here $(BOD_u)_0$ is the (initial) ultimate BOD and $(BOD_u)_n$ the (ultimate) BOD remaining after n d; thus their difference is the BOD satisfied or removed in n d (Figure 4.1). Since bacteria cannot, by definition, oxidize non-biodegradable compounds, $(NBDG)_0 = (NBDG)_n$. Thus:

$$\Delta COD_n = BOD_n$$

A suitable procedure is:

(1) Measure initial COD of the waste sample.
(2) Place 50–100 ml of the sample in a 500 ml conical flask, stopper with *non*-absorbent cottonwool and place in a cooled incubator at 20 °C. Incubate for n d (usually $n = 5$). Sufficient oxygenation of the sample is provided through the large surface area of the small sample volume in the large flask. Evaporation losses are minimized if a large shallow tray of water is placed on the incubator floor.
(3) After n d make up any evaporation losses with distilled water and measure the COD remaining and hence calculate the BOD_n.
(4) If a cooled incubator is not available BOD_5 may be estimated as the $2\frac{1}{2}$ d BOD at 35 °C (Section 4.5, example 2).
(5) If the BOD curve is required, measure the COD remaining each day. (In this case a larger sample is necessary, e.g. 500 ml sample in a 1000 ml flask; use a magnetic stirrer to ensure adequate oxygen transfer.)

Sample preservation[9]

At ambient temperatures > 20 °C samples must be analysed within 12 h and at temperatures of 10–20 °C within 24 h. Samples may however be stored at 1 °C for 6–7 d. No preservative should be added to the sample (it would kill the essential bacteria).

Algae

Stabilization pond effluent and other samples containing algae should be incubated in the *dark*. If they are incubated in the light, oxygen release through photosynthesis occurs and there will be little or no DO depletion in the BOD bottle—indeed there may be a net DO gain.

Nitrification

The oxygen demand exerted during the oxidation of ammonia to nitrate (Section 3.3) can be a very high proportion of the total BOD of an effluent. For nitrification to occur both ammonia and nitrifying bacteria must be present. In raw sewage, for example, only the former is present and in a fully nitrified effluent only the latter; thus in both cases the nitrification BOD would be zero. In *partially nitrified effluents* both are present and rapid nitrification can occur right from the start of the BOD test (Figure A1.2)

If the BOD of a final effluent is being measured to determine the effect of the effluent on the receiving stream, then nitrification should be allowed to proceed in the BOD bottle. In this way the BOD test estimates the maximum

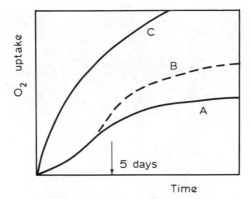

Figure A1.2 BOD curves showing the effect of nitrification. Curve A shows the wholly carbonaceous demand of a fully nitrified effluent. Curve B shows the nitrification demand of a non-nitrified effluent which becomes apparent only after 5–6 d. Curve C shows the combined carbonaceous and nitrification demand of a partially nitrified effluent

oxygen demand that would be exerted in the receiving stream. Yet if the stream temperature is < 20 °C, this may be an overestimate as the rate of nitrification falls rapidly with temperature.[10] However, if the BOD is to be used as a measure of the biodegradable organic compounds present, then it is preferable to prevent nitrification occuring during the test so that only the 'carbonaceous' BOD may be measured. A suitable nitrification inhibitor is allythiourea: 2 ml of a 250 mg/l solution should be added to each litre of the dilution water or the sample dosed at a rate of 0·5 mg/l. Regulatory agencies should be encouraged to set separate effluent standards for carbonaceous BOD_5 and NH_3. In this way any possible confusion between different interpretations of a simple BOD_5 standard would be avoided.

Faecal coliform counts

FC counts > 10^4/100 ml may be estimated by the agar dip-slide technique.[11] An agar dip-slide[12] (Figure A1.3) is immersed in the stream or wastewater flow for 5–10 seconds and replaced in its container. After incubation at 44 °C for 24 h the count may be estimated to the nearest order of magnitude from the chart shown in Figure A1.4. Total coliform counts may be obtained by incubation at 35–37 °C for 24 h.

For counts < 10^4/100 ml a Millipore[13] dip-slide kit is available. This kit is held in the flow for 30 seconds during which time exactly 1 ml is sucked through a membrane filter into an absorbent pad immediately behind the

150

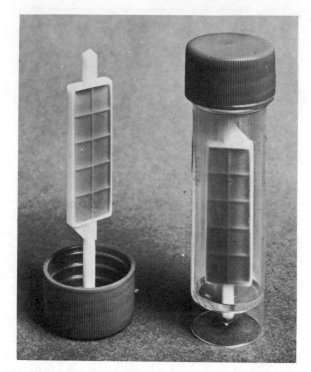

Figure A1.3 Agar dip-slide. Some of the bacteria in the sample adhere to the agar jelly which contains an abundance of bacterial food. During incubation each bacterium on the agar surface multiplies to form a *colony*. The approximate number of bacteria on the sample is obtained by comparing the colony density with the chart shown in Figure A1.4

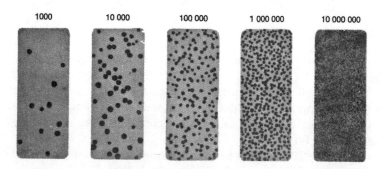

Figure A1.4 Colony comparison chart for use with agar dip-slides
[Courtesy of Orion Pharmaceutical Ltd]

filter. The membrane filter is a very fine filter paper with a maximum pore size < 0.5 μm; all bacteria are retained on its surface. The absorbent pad contains dehydrated FC broth, a special medium which permits only the growth of FC. The number of blue colonies appearing after 24 h at 44 °C are counted and the result multiplied by 100 to give the normal FC count. For FC levels $< 10^2/100$ ml in fairly turbid waters (such as sewage effluent) the standard membrane filtration procedure should be used; portable field kits are available.[13]

Other Tests

Relative stability

The stability of an effluent is defined as its ability to maintain its aerobic condition for 5 d when incubated at 20 °C in the absence of air. An unstable effluent is likely to cause nuisance in the receiving watercourse. Methylene blue is added to the sample as an indicator of aerobiosis; under anaerobic conditions it is rapidly reduced to the leuco (colourless) base. This test is particularly suitable for use at small works. A suitable procedure is:

(1) Pipette 1 ml of methylene blue solution (0·35 g/l) into a 250–300 ml BOD bottle. Slowly fill the bottle with the waste, avoiding aeration as far as possible. Stopper so as to exclude any bubbles of air. Incubate at 20 °C.
(2) Examine the bottle after $\frac{1}{2}$, 1, 2 and 4 h and then twice daily. The effluent has failed the test if, before the end of 5 d, the methylene blue is reduced to its leuco base. Record the time at which failure was noticed.

> There are charts available which give the 'percentage stability' corresponding to the time of failure. These charts are based on equation 4·2 with $k_1 = 0.23$ d^{-1} ($= 0.1$ d^{-1}, base 10). The evidence for proper use of this equation in stability problems is sparse and it appears more realistic to report only whether the effluent has passed the test or not and, if it has not, the time at which failure occurred.

Suspended solids

The standard procedure is filtration through a glass-fibre filter paper.[13] A rough estimate of whether an effluent has a SS concentration < 30 mg/l can be made by measuring its *transparency*.[14] The required apparatus is a clear glass or plastic tube 600 mm long, 25 mm dia., which has a flat base. The tube is calibrated every 5 or 10 mm from 0 to 600 mm. On the outside of the base is stuck a disc of white paper having a central black cross with lines 1 mm thick. The sample is poured into the tube until the black cross just disappears from view; the depth of liquid is recorded. If the SS is < 30 mg/l the transparency will be > 300 mm. (This test was developed for conventional effluents and is thus applicable to the settled effluent from aerated lagoons and oxidation ditches; so far it has not been much used with stabilization pond effluents.)

Ammonia[1,14]

Clarify the sample by adding 1 ml zinc sulphate solution (100 g/l $ZnSO_4.7 H_2O$) to 100 ml. Mix and then add 0·4–0·5 ml 6N sodium hydroxide solution (240 g/l); mix and allow to stand for a few minutes. A dense precipitate will form and settle to the bottom. Pipette 1–2 ml of the clarified effluent into a 50 ml Nessler cylinder, add about 20–30 ml distilled water (ammonia free) and 1 ml of 10% sodium hexametaphosphate solution (to prevent the precipitation of magnesium hydroxide); mix and make up to 50 ml with distilled water. Add 2 ml BDH Nessler's reagent and match the colour produced with a standard glass colour disc NAD in a BDH Nesslerizer.[15] Alternatively 0·1–0·5 ml of the clarified sample may be tested in the Hach NI-8 ammonia test kit.[6]

Sampling

Because sewage varies in both magnitude of flow and strength throughout the day (Figure 3.1), a single 'grab' sample does not provide much information. As we usually wish to know the mean daily characteristics of a wastewater, we must obtain a *flow-weighted composite sample*. This means samples have to be taken every 2 h and mixed together in proportion to the flow at the time of their collection. For example:

Time	Flow (m^3/d)	Volume of sample added to composite (ml)
0800	1150	115
1000	1420	140
1200	1760	175
1400	1380	140
1600	990	100
etc.		

REFERENCES

1. *Standard Methods for the Examination of Waters and Wastewaters*, 13th edition, American Public Health Association, New York, 1971.
2. Gaudy, A. F. and Ramanathan, M., *Journal of the Water Pollution Control Federation*, **36**, 1479 (1964).
3. Jobling Laboratory Division, Stone, Staffordshire, UK.
4. Walden Precision Apparatus Ltd, Shire Hill, Saffron Walden, Essex, UK.
5. Masselli, J. W. *et al.*, *Water Pollution Control Deeds and Data*, (6), D7 (1971).
6. Hach Chemical Company, PO Box 905, Ames, Iowa 50010, USA.
7. Montgomery, H.A.C. *et al.*, *Journal of Applied Chemistry*, **14**, 280 (1964).
8. Gaudy, A. F. and Gaudy, E. T., *Industrial Water Engineering*, **9** (5), 30 (1972).
9. Loehr, R. C. and Bergeron, B., *Water Research*, **1**, 577 (1967).
10. *Notes on Water Pollution No. 52*, HMSO, London, 1971.
11. Mara, D. D., *Water Research*, **6**, 1605 (1972).
12. Oxoid Ltd, Southwark Bridge Road, London, SE1 9HF, UK; Orion Pharmaceutical

Co., PO Box 10019, Helsinki 10, Finland; Stayne Laboratories Ltd, Marlow, Buckinghamshire, UK.
13. Millipore Intertech Inc., PO Box 255, Bedford, Mass 01730, USA. (Application procedure sheet No. PB407 for dip-slide kit, No. AB313 for standard FC membrane filter count, No. AP 312 for SS analysis.)
14. *Notes on Water Pollution No. 44*, HMSO, London, 1969.
15. BDH Chemicals Ltd, Poole, Dorset BH12 2NN, UK.

Further reading

C. N. Sawyer and P. L. McCarty, *Chemistry for Sanitary Engineers*, 2nd edition, McGraw-Hill, New York, 1967.
Simple Methods of Testing Sewage Effluents (Notes on Water Pollution No. 44) HMSO, London, 1969.

Appendix 2

Analysis of BOD Data

First Order Kinetics

Here we have a series of n experimental results of BOD removal with time which are to be fitted to equation 4.4 in order to obtain estimates of the values of k_1 and L_0:

$$y = L_0(1 - e^{-k_1 t})$$

Standard computer programs are available for the analysis of this type of equation by the method of least squares. In the absence of computer facilities the following 'manual' least squares analysis may be used.

Combine equations 4.1 and 4.3 to yield:

$$d(L_0 - y)/dt = -k_1(L_0 - y)$$

i.e.

$$dy/dt = k_1(L_0 - y) \tag{1}$$

For each data point (t_i, y_i) this equation may be written as:

$$(dy/dt)_i = k_1(L_0 - y_i) \tag{2}$$

Now owing to experimental error the two sides of this equation will not be equal but will differ by a residual amount R_i:

$$R_i = k_1(L_0 - y_i) - (dy/dt)_i \tag{3}$$

Writing $(dy/dt)_i$ as y_i' and substituting p for $k_1 L_0$ and q for k_1, we have:

$$R_i = p - qy_i - y_i' \tag{4}$$

For the sum of the squares of the residuals to be a minimum:

$$\frac{\partial}{\partial p}\sum(R_i)^2 = \frac{\partial}{\partial q}\sum(R_i)^2 = 0$$

i.e.

$$\sum 2R_i(\partial R_i/\partial p) = 0$$

and

$$\sum 2 R_i(\partial R_i/\partial q) = 0$$

Applying these conditional equations to equation 4, we obtain the following pair of simultaneous equations in p and q:

$$np - q\sum y_i - \sum y_i' = 0 \tag{5}$$

$$-p\sum y + q\sum (y_i)^2 + \sum (y_i y_i') = 0 \tag{6}$$

The following approximation is used for y_i' (Figure A2.1):

$$y_i' = \frac{y_{i+1} - y_{i-1}}{2\,\Delta t} \tag{7}$$

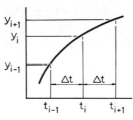

Figure A2.1

Example. Calculate k_1 and L_0 for the following data:

t_i	1	2	3	4	5	6
y_i	8	11	15	18	20	21

Solution

Calculate y_i^2, y_i' and $y_i y_i'$ for each data point and so obtain $\sum y_i$, $\sum y_i^2$, $\sum y_i'$ and $\sum y_i y_i'$:

t	y_i	y_i^2	y_i'	$y_i y_i'$
1	7	49	5·5	38·5
2	11	121	4·0	44·0
3	15	225	3·5	52·5
4	18	324	2·5	45·0
5	20	400	1·5	30·0
	71	1119	17·0	210·0

Note that the last data point (here t_6, y_6) is used only to calculate the penultimate y_i' (here y_5'); also that y_1' is calculated with $y_0' = 0$.

Substitution into equations 5 and 6 yields:

$$5p - 71q - 17 = 0$$

$$-71p + 1119q + 210 = 0$$

156

From which $p = 7 \cdot 4$ and $q = 0 \cdot 28$, so that:

$$k_1 = 0 \cdot 28 \text{ d}^{-1}$$
$$L_0 = 26 \text{ mg/l}$$

Second Order Kinetics

Equation 4.15 can be written as:

$$\frac{L_0 - y}{L_0} = \frac{1}{1 + k_2 L_0 t} \tag{8}$$

which can be rewritten in linear form as:

$$\frac{1}{y} = \left(\frac{1}{k_2 L_0}\right)\frac{1}{t} + \frac{1}{L_0} \tag{9}$$

Thus a plot of y_i^{-1} against t_i^{-1} should yield a straight line of slope $(k_2 L_0)^{-1}$ and intercept $(L_0)^{-1}$.

Appendix 3

Elements of Sanitary Sewer Design

Elements of Hydraulic Design

The flow of sewage in sewers is, wherever possible, gravitational and at atmospheric pressure. Sewers are designed to run only partially full in normal operation; they are thus designed as open channels. The most commonly used design equation is the Manning formula:

$$v = \frac{1}{n}m^{2/3}i^{1/2} \tag{1}$$

where v = velocity of flow, m/s
n = roughness coefficient
m = hydraulic mean depth (area of flow ÷ wetted perimeter), m
i = hydraulic gradient (pipe slope).

For concrete pipes (which are in the most common use) $n = 0.012–0.015$, with 0.013 being the usual design value. For PVC pipes (which are being increasingly used for small sewers up to 300 mm dia.) $n = 0.009–0.011$. The capacity, Q, of the sewer is given by:

$$Q = \frac{A}{n}m^{2/3}i^{1/2} \tag{2}$$

where A = cross-sectional area of flow, m².
Sanitary sewers are designed to carry $4 \times DWF$ when running just full. This provides an adequate factor of safety above the normal daily peak flow of $1\frac{1}{2} - 2\frac{1}{2} \times DWF$ (Section 3.2). (In fact, due to the properties of circular sections (Figure A3.1), the maximum discharge is some 7 per cent higher, about $4.3 \times DWF$.)

Self-cleansing velocities

The velocity of flow must be high enough to prevent the deposition of sewage solids which would otherwise cause blockage. A velocity of at least 0.6 m/s is necessary, and 1 m/s is preferable. It is usual to ensure that the velocity does not fall below this value when the sewage is flowing at its DWF.

If the self-cleansing velocity is taken as 1 m/s at DWF and the 'just full' flow is $4 \times DWF$, then, from Figure A3.1:

158

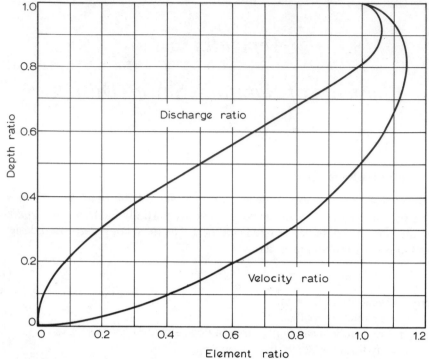

Figure A3.1 Flow in circular sewers. The depth (or flow, velocity) ratio is the ratio of the depth of flow (or flow, velocity of flow) to the sewer diameter (or flow and velocity when just full). No correction has been made for variations in roughness (Manning's n) with depth

(1) At DWF the ratio $q/Q = 0.25$ and the corresponding velocity ratio $v/V = 0.84$. Therefore if the velocity at DWF is 1 m/s the 'just full' velocity $= 1/0.84, = 1.2$ m/s.

(2) The velocity of flow will be > 0.6 m/s (i.e. $v/V > 0.5$) for $q/Q > 0.05$ — i.e. for flows $> 0.2 \times$ DWF.

These are *minimum* figures and they are often exceeded. A minimum self-cleansing velocity means in effect that there is a minimum slope for each size of pipe. Minimum slopes for various sizes of concrete pipe are given in Table A3.1. PVC pipes, being smoother, can be laid at flatter gradients but in practice seldom are.

Minimum sewer sizes

To reduce the frequency of blockages public sewers should not be smaller than 200–225 mm, even though the hydraulic calculations may indicate that a smaller size would be satisfactory. House connections should be 100–150 mm dia.

Table A3.1 Minimum sewer gradients

Diameter* (mm)	Minimum gradient m/km
225	11
300	8
375	6
450	$4\frac{1}{2}$
600	3
750	$2\frac{1}{4}$
900	$1\frac{3}{4}$
1050	$1\frac{1}{2}$
1200	$1\frac{1}{4}$

*Other diameters are available as standard.
[†]Based on equation 1 with $n = 0.013$ and $V = 1.2$ m/s (equivalent to 1 m/s at DWF for sewers designed to carry $4 \times$ DWF). For a velocity of 0.7 m/s (equivalent to 0.59 m/s at DWF) these gradients should be multiplied by $(0.7/1.2)^2, = 0.34$.

Large sewers

In small sewers of less than 1200 mm dia the flow is usually *uniform*, i.e. the gradient of the water surface is approximately the same as the gradient of the sewer invert. Thus in the Manning formula the hydraulic gradient i is taken to be the pipe gradient. In larger sewers the flow is often *non-uniform* and due allowance must be made for this condition.

Manholes

Manholes are inspection boxes which provide access from ground level for the removal of blockages by rodding. They should be provided at every change of direction, size and slope; on straight runs they should be located at least every 80–100 m for sewers up to 1200 mm dia. and every 300–400 m for larger sizes.

The Zambian Sewerage System

The high capital costs of a sewerage scheme can be considerably reduced by minimizing the amount of trench excavation required. If settlement chambers are located along each branch of the sewerage network, the downstream sewage flow does not contain any solids; self-cleansing velocities are not therefore required and the sewers can be laid at very small gradients to provide only nominal velocities of 0.2–0.3 m/s at full flow. In Zambia, where most sewered areas are on very flat ground, aqua-privies are used as the settlement chamber;[1] alternatively septic tanks may be used.[2] This system of sewerage requires more maintenance (mainly sludge removal) than does the usual system, but it has the advantages not only of lower overall costs but also of being slightly less capital intensive and rather more labour intensive (see Section 14.1).

REFERENCES

1. Marais, G. v. R., in *Progress in Water Technology*, Vol. 3, Pergamon, Oxford, 1973.
2. Martin, A. J., *The Work of the Sanitary Engineer*, MacDonald & Evans, London, 1935, p. 199.

Further reading

R. E. Bartlett, *Sewerage: Design in Metric*, Applied Science Publishers, London, 1970.
J. B. White, *The Design of Sewers and Sewage Treatment Works*, Edward Arnold, London, 1970.

Appendix 4

Procedures for Facultative Pond Design

In addition to the design procedures described in Section 7.9 there are several more in common use. Some of these are discussed below. It will be noticed however that none of these can be recommended for general use in hot climates.

The Gloyna Procedure

Hermann and Gloyna[6] working in Texas found that the optimum temperature for pond operation was 35 °C. The retention time t_T^* required at any temperature T for 80–90 per cent BOD removal from a typical domestic sewage in USA ($BOD_5 = 200$ mg/l) was shown to be related to the retention time at 35 °C by the following Arrhenius equation:

$$t_T^* = t_{35}^* \theta^{35-T} \tag{1}$$

so that the mid-depth area A was given by:

$$A = Qt_T^*/D \tag{2}$$

where Q = flow, m^3/d
   D = depth, m.
Therefore:

$$A = t_{35}^* \theta^{35-T}(Q/D) \tag{3}$$

For wastes with a BOD_5 (L_i) other than 200 mg/l, the ratio $L_i/200$ was introduced:

$$A = t_{35}^* \theta^{35-T}(Q/D)(L_i/200) \tag{4}$$

Practical use of these equations requires, of course, knowledge of the values of t_{35}^* and θ. The following values have been reported or used:

for t_{35}^2: 3·5 d (Hermann and Gloyna[6])
       7·5 d (Marais[14])
       7·0 d (Huang and Gloyna[7]; Gloyna[5])
for θ : 1·072 (Hermann and Gloyna[6])
       1·085 (Marais[14]; Gloyna[5])

The large variation in the values for t_{35}^* is the major factor which militates against the use of this procedure, since for any given combination of L_i, Q and T, its value controls the size of the pond and hence its cost. Indeed the original choice of 35 °C as the reference temperature was somewhat unfortunate as field data at this temperature are extremely few. Since data interpolation is considerably less inaccurate than extrapolation, 15 °C or 20 °C would have been a more useful choice. Some other points are:[12]

(1) Hermann and Gloyna[6] state that the pond volume derived from their version of equation 4 was the overall volume of a series of two, preferably three, ponds. In constrast Huang and Gloyna[7] state that their version (which differs only in its values for t_{35}^* and θ) gives the volume of a single pond of depth 1·83 m.

(2) Gloyna[4] states that the ratio $L_i/200$ is valid only when the value of L_i is a 'proximal deviation' from 200 mg/l. The term 'proximal deviation' was not defined but we may assume that the procedure is unsuitable for $L_i > 300$ mg/l.

(3) Gloyna[5] stipulates that L_i should be taken as the 5-day BOD for 'weak or presettled' wastewaters but as the ultimate BOD for 'strong or untreated' wastewaters. No reason for this recommendation was given.

Thus it appears that the Gloyna procedure is not suitable for use in hot climates.

Surface Load Procedures

Design procedures based wholly or in part on the daily areal BOD_5 loading are commonly used by design engineers. The simplest of these procedures is purely empirical in which the mid-depth area is calculated from equation 7.7:

$$A = \frac{10\,QL_i}{\lambda_s}$$

where λ_s = design BOD_5 loading, kg/ha d.

The value of λ_s is chosen on the basis of experience of pond performance in the local or a similar climate; for example the first ponds built in Kenya were designed on the basis of experience gained in South Africa with $\lambda_s = 200$ lb/acre d.[8] More frequently however the design value of λ_s is stipulated by the appropriate regulatory agency; for example most states in USA have design criteria based on organic loading and minimum detention time.[2]

The simplicity of this empirical procedure has much to recommend it, always providing of course that there are enough local field data of sufficient quality which can be used to determine the most suitable design value for λ_s. In regions with little or no experience of waste stabilization ponds, the McGarry and Pescod procedure should be adopted.

Procedures based on solar energy

These procedures, the first of which was developed by Oswald and Gotaas[18] for high-rate aerobic ponds, are surface load procedures based on equating the daily BOD removal in the pond with the daily production of oxygen by the pond algae which is in turn based on the intensity of solar radiation received at the pond surface (this latter quantity is measured in langleys ($=$ cal/cm^2 d) and is denoted by the symbol Λ).

Algae utilize only about 6 per cent of the incident solar energy and about 6 million calories are required to produce 1 kg of algae. Therefore the weight W (kg/ha d) of algae produced in the pond is given by:

$$\frac{\text{langleys} \times 10^8 \text{ (cal/ha d)} \times 0{\cdot}06 \text{ (efficiency)}}{6 \times 10^6 \text{ (cal/kg)}}$$

Numerically therefore $W = \Lambda$. Now approximately 1·6 kg of oxygen are produced photosynthetically for every 1 kg of algae growth, so that the rate of oxygen production O (kg/ha d) in the pond is given by:

$$O = 1{\cdot}6\,\Lambda \tag{5}$$

The oxygen production is equated with the removal of ultimate BOD in the lagoon.[1,9] A slightly more conservative criterion[18] is to equate the oxygen production with the ultimate BOD applied (rather than removed) per ha per day, λ_{us}:

$$\lambda_{us} = 1{\cdot}6\Lambda \tag{6}$$

Since the ultimate BOD $= ca.$ 1·5 $\times$ BOD$_5$, equation 6 can be written:

$$\lambda_s = 1{\cdot}07\Lambda \tag{7}$$

The mid-depth area is now calculated from equation 7.7.

Equations 6 and 7 are unsatisfactory since they imply that all the oxygen produced by the algae is available for stabilization of the influent waste. This is not necessarily true and a suitable modification to equation 7 is:

$$\lambda_s = \beta\Lambda \tag{8}$$

where $\beta < 1{\cdot}07$ and is a factor of safety which includes the effect of the un-availability of algal oxygen for waste stabilization.

There are conflicting reports on the value of β: in India 0·5–1·0 langley was found to be required to stabilize 1 lb/acre d (1·12 kg/ha d) whereas in the USA 1·5–2·0 langleys were found necessary.[3,10] Thus design procedures based on incident solar energy cannot be recommended for general use since the 'constant' β appears to vary from region to region.

Marais and Shaw Procedure[16]

Thus is a procedure based on first order kinetics and the assumption of complete mixing. It differs from the procedure given in Section 7.9 only in that the following empirical equation (based on field data from southern African and southern USA) is used to relate the maximum pond BOD$_5$ consistent with the maintenance of predominantly aerobic conditions to the pond depth:

$$L_e = \frac{N}{2D + 8} \tag{9}$$

where N = a constant

[here D is in m; the original equation, with D in ft, was $L_e = N/(0.6D + 8)$].
The original field data indicated that $N = 1000$, but for the purposes of design
this was reduced to:

750	(Marais and Shaw[16])
700	(Marais[15])
600	(Meiring et al.[17])

Equation 9 indicates that the variation of L_e with depth is very small indeed,
at least within the normal range of depths (1–1.5 m), for whatever value of N
is chosen (Table A4.1). Equation 9 really shows that L_e is more properly considered *independent* of D for $1 < D < 1.5$ and that designers may as well choose
a value for L_e directly, rather than indirectly via some choice of value for N.[11,12]

Table A4.1 Design values of L for various pond
depths*

D	L_e (mg/l)			
(m)	$N = 1000$	$N = 750$	$N = 700$	$N = 600$
1.0	100	75	70	60
1.1	98	74	69	59
1.2	96	72	67	58
1.3	94	71	66	57
1.4	93	69	65	56
1.5	91	68	64	55

*Calculated from equation 9.

Value of k_1

The selection of a suitable value of k_1 is the most difficult part of using design
procedures based on first order kinetics. In southern Africa the value of k_1
was found to be 0.23 d^{-1} which Marais and Shaw[16] reduced for design purposes
to 0.17 d^{-1}. Presently recommended practice in South Africa is to choose
k_1 as 0.17 d^{-1} if the average temperature during the coldest month is > 5 °C,
or 0.14 d^{-1} if it is < 5 °C.[17] Marais[14] modified equation 4.7 by permitting k_1
to vary with temperature according to an Arrhenius equation. Because Marais
was attempting to integrate the Hermann and Gloyna and the Marais and Shaw
theories of BOD$_5$ removal in ponds, the reference temperature was chosen
as 35 °C:

$$k_T = k_{35}\theta^{T-20} \tag{10}$$

Marais analysed the results which Suwannakan[19] obtained from laboratory model ponds fed with a synthetic milk waste and found that $k_{35} = 1 \cdot 2 \, d^{-1}$ and $\theta = 1 \cdot 085$. These values are not known to be applicable to domestic sewage and they cannot therefore be used with confidence for the design of full-scale ponds. Mara[13] suggested that a suitable design equation for k_T was equation 7.12:

$$k_T = 0 \cdot 30 (1 \cdot 05)^{T - 20}$$

Here the reference temperature is 20 °C and the design value of k_{20} is conservatively taken as $0 \cdot 30 \, d^{-1}$. Equation 7.12 should only be used if $T > 15$ °C, otherwise the value of the Arrhenius constant may be too small. Zanker[21] suggested $\theta = 1 \cdot 054$ and $k_{20} = 0 \cdot 80 \, d^{-1}$; the latter value seems too high for general use, although no doubt it reflects experience gained in Israel[20] where a BOD_5 reduction from 250 mg/l to 50 mg/l in 5 d is a normal design assumption.

REFERENCES

1. Arceivala, S. J. et al., Waste Stabilization Ponds; Design, Construction and Operation in India, Central Public Health Engineering Research Institute, Nagpur, 1970.
2. Canter, L. W. and Englande, A. J., Journal of the Water Pollution Control Federation, 42, 1840 (1970).
3. Clare, H. C. et al., in Proceedings of a Symposium on Waste Stabilization Lagoons, Kansas City, 1960, US Public Health Service, Washington, DC, 1961.
4. Gloyna, E. F., in Advances in Water Quality Improvement, University of Texas Press, Austin, 1968.
5. Gloyna, E. F., Waste Stabilization Ponds, World Health Organization, Geneva, 1971.
6. Hermann, E. R. and Gloyna, E. F. Sewage and Industrial Wastes, 30, 963 (1958).
7. Huang, J-C. and Gloyna, E. F., Water Research, 2, 459 (1968)
8. Hunt, M. A. and Westenberg, H. J., Journal and Proceedings of the Institute of Sewage Purification, (3), 230 (1964).
9. Jayangoudar, I. S. et al., Journal of the Water Pollution Control Federation, 42, 1501 (1970).
10. Lakshminarayana, J. S. et al., in Proceedings of a Symposium on Waste Treatment by Oxidation Ponds, 1963, Central Public Health Engineering Research Institute, Nagpur, 1964.
11. Mara, D. D., Water Research, 8, 493 (1974).
12. Mara, D. D., Water Research, 9, 595 (1975).
13. Mara, D. D., Design Manual for Sewage Lagoons in the Tropics, E. A. Literature Bureau, Nairobi, 1975.
14. Marais, G. v. R., Bulletin of the World Health Organization, 34, 737 (1966).
15. Marais, G. v. R., in Proceedings of the Second International Symposium on Waste Treatment Lagoons, Kansas City, University of Kansas, Lawrence, 1970.
16. Marais, G. v. R. and Shaw, V. A., Transactions of the South African Institution of Civil Engineers, 3, 205 (1961).
17. Meiring, P. C. J. et al., CSIR Special Report WAT 34, Centre for Scientific and Industrial Research, Pretoria, 1968.
18. Oswald, W. J. and Gotaas, H. B., Transactions of the American Society of Civil Engineers, 122, 73 (1957).
19. Suwannaken, V., Temperature Effects on Waste Stabilization Ponds, Ph.D. thesis, University of Texas, 1963.

20. Watson, J. L. A., *Proceedings of the Institution of Civil Engineers*, **22**, 21 (1962).
21. Zanker, A., *Water Pollution Control Federation Deeds & Data*, (10), D2 (1973).

*In spite of its apparent simplicity
the waste stabilization pond is
a complex biochemical reactor
which defies precise design.*[1]

Index

Marine pollution, 32
Maturation ponds, 74, 84, 118
McGarry and Pescod design procedure, 82
Microbes, 8
Microstrainers, 116
Mid-depth area, 80
Mineralization, 24
Mixing, 71
Muskegon irrigation scheme, 133

Nightsoil collection, 237
    treatment, 139
Nitrification, 148
Non-faecal coliforms, 13
Non-waterborne systems, 7
Nutrients, 25
Nutrification, 26

Odour, 67, 86
Oxidation ditches, 103
Oxygen sag curve, 28

Parasites, 20
Pathogenic bacteria, 11
Pebble bed clarifier, 117
Percolation tests, 123
Photosynthesis, 17
Plastic media, 59
Plug flow, 37
Pond facilities, 86
Preliminary treatment, 44, 53
Protein, 77, 127
Protozoa, 17

Recirculation, 58
Reclamation, 127, 134
Relative stability, 151
Retarded exponential, 40
Re-use, 127
Rock filter, 118
Royal Commission standard, 29

Sampling, 152
Sand filtration, 117
Schistosomiasis, 27
Screening, 44
Second order kinetics, 40, 154

Sedimentation, 54
Self-purification, 28
Septic tanks, 120
Sewage, 1
    collection, 7, 21
    flows, 23
    strength, 4, 23
    treatment, 22
    treatment costs, 76
    treatment works location, 22
Sewers, 7
    design, 157
    sizes, 158
Sludge layer, 72
    loading factor, 105
    treatment, 65
Sodium absorption ratio, 130
Solar energy design procedure, 162
Soluble BOD, 96
Stepped aeration, 63
Stormwater, 24
Stratification, 72
Subsurface irrigation, 122
Sullage, 1
Surface load procedure, 162
Surface water pollution, 27
Suspended solids measurement, 151
Symbiosis, 70

Temperature, 9
Tertiary treatment, 115
Theoretical oxygen demand, 3
Thermocline, 72
Trickling filter, 57

Upflow filters, 123
Urine, 1

Viruses, 17

Waste stabilization ponds, 69
Wastewater management, 21
Worms, 20

Yield coefficient, 96

Zambian sewerage system, 159

ٮ